Exceptional Technologies

ALSO AVAILABLE FROM BLOOMSBURY

Philosophy and Simulation, Manuel DeLanda
Speculative Realism, Peter Gratton
Epistemology of Noise, Cecile Malaspina
Retroactivity in Contemporary Art, Craig Staff

Exceptional Technologies: A Continental Philosophy of Technology

DOMINIC SMITH

BLOOMSBURY ACADEMIC
LONDON • NEW YORK • OXFORD • NEW DELHI • SYDNEY

BLOOMSBURY ACADEMIC
Bloomsbury Publishing Plc
50 Bedford Square, London, WC1B 3DP, UK

BLOOMSBURY, BLOOMSBURY ACADEMIC and the Diana logo are trademarks of
Bloomsbury Publishing Plc

First published in Great Britain 2018

Copyright © Dominic Smith, 2018

Dominic Smith has asserted his right under the Copyright, Designs and Patents Act, 1988,
to be identified as Author of this work.

For legal purposes the Acknowledgements on p. x constitute an extension of this
copyright page.

Cover design: Irene Martinez-Costa

Cover image © David Evan Mackay

All rights reserved. No part of this publication may be reproduced or transmitted
in any form or by any means, electronic or mechanical, including photocopying,
recording, or any information storage or retrieval system, without prior
permission in writing from the publishers.

Bloomsbury Publishing Plc does not have any control over, or responsibility for, any
third-party websites referred to or in this book. All internet addresses given in this
book were correct at the time of going to press. The author and publisher regret any
inconvenience caused if addresses have changed or sites have ceased to exist,
but can accept no responsibility for any such changes.

A catalogue record for this book is available from the British Library.

A catalog record for this book is available from the Library of Congress.

ISBN: HB: 978-1-3500-1560-9
PB: 978-1-3500-1561-6
ePDF: 978-1-3500-1559-3
eBook: 978-1-3500-1562-3

Typeset by Deanta Global Publishing Services, Chennai, India

To find out more about our authors and books visit www.bloomsbury.com and
sign up for our newsletters.

To Joanna and Orrin, with love

CONTENTS

List of Figures ix
Acknowledgements x

Introduction: Picturing technology 1
   1 Picturing technology today 1
   2 Key terms 5
   3 Structure and limits 7

1 A sense of the transcendental 11
   1 Malabou's sense 12
   2 Expanding sense 16
   3 Expanding further: From minimal to maximal sense 23
   4 Philosophy of technology: Making sense of many turns 27

2 The blank page 35
   1 'This white paper' 37
   2 Varying conditions 42
   3 Re-imagining relevance (1) 45
   4 Re-imagining relevance (2) 51

3 Embodiment conditions 55
   1 *On the Internet* 57
   2 A developing body of work 62
   3 Situating embodiment conditions: 4e 69
   4 Crossover potentials: Between philosophy of technology, media theory and 4e 73

4 Three exceptional technologies 77

    1   Everything but the network: Vannevar Bush's Memex 79
    2   'Pictorial statistics': Francis Galton's composite photography 86
    3   'Machine with concrete': Arthur Ganson's gestural engineering 97
    4   Problems and prospects 104

5 Which way to turn? 107

    1   The empirical turn: An enduring influence in philosophy of technology? 109
    2   The speculative turn: A new beginning in continental philosophy? 114
    3   An alternative picture: Method as 'Mapping' 119
    4   A shared field of exceptional complexities 124

Conclusion: Exceptional technologies, not technological exceptionalism 129

Notes 133
References 151
Index 166

LIST OF FIGURES

Figure 1   The 'memex', based on Bush's description in 'As We May Think'. Courtesy Elsevier. This image was published in *From Memex to Hypertext*, Nyce, J. M. and P. Kahn (eds), 'As We May Think', p 109, Copyright Elsevier (1991)   80

Figure 2   Examples of composite photographs from Galton's *Inquiries into Human Faculty* (1883). Galton, F. (1907), *Inquiries into Human Faculty and Its Development*, 2nd edn, London: Dent, p 8   87

Figure 3   Galton's basic wooden rig, c. 1878. Galton, F. (1907), *Inquiries into Human Faculty and Its Development*, second edition, London: Dent, p 223   88

Figure 4   Galton's more specialized apparatus, c. 1881. Galton, F. (1907), *Inquiries into Human Faculty and Its Development*, second edition, London: Dent, p 222   88

Figure 5   Detail of Galton's brass framing grid, c. 1878. Galton, F. (1907), *Inquiries into Human Faculty and Its Development*, second edition, London: Dent, p 235   89

Figure 6   Bionomial distribution. Quetelet, A. (1846). *Lettres sur la théorie des probabilités, appliquée aux sciences morales et politiques*, Brussels: M. Hayez, p 103   95

Figure 7   Arthur Ganson's 'Machine with Concrete'. Courtesy Amy Snyder. Copyright Exploratorium, www.exploratorium.edu   98

ACKNOWLEDGEMENTS

Many different people helped me work towards this book. Colleagues past and present at the University of Dundee helped me find the time and confidence to complete the book (Nicholas Davey, Beth Lord, James Williams, Ashley Woodward, Tina Röck, Oisín Keohane, Brian Smith, Todd Mei and Patrick Levy). Walter Pedriali, James Williams and Brian Smith provided invaluable feedback on early drafts, and participation in the *Sujet Digital* research group at Paris VIII helped me develop key ideas (special thanks to Pierre Cassou-Noguès, Claire Larsonneur, Arnaud Regnauld, Sara Touiza Ambrogianni, Erika Fülöp and Gwen Le Cor). Thank you also to Mark Coeckelbergh, Galit Wellner, Pieter Lemmens, Sarah Cook and Yoni Van Den Eede for providing valuable moments of encouragement.

Thank you to the School of Humanities at the University of Dundee for allowing me a research focused semester to work towards completion of the book. Thank you also to Frankie Mace and Liza Thompson at Bloomsbury for their excellent editorial support, and to David Evan Mackay for use of the cover image. Thank you also to Ben Whitney, who first made me aware of the work of Arthur Ganson.

An earlier version of material included in Chapter 2 appeared in a 2015 collection (Larsonneur, C., Regnauld, A., Cassou-Noguès, P., Touiza, S., (eds) (2015), *Le sujet digital*, Dijon: Les Presses du réel). I gratefully acknowledge permission to reprint this material.

I would like to extend three words of special thanks. First, to all the students I have worked with at the University of Dundee: it has been/continues to be an immense privilege to work with you. Second, my friends and family have been an invaluable source of support and encouragement (especially Mum, Kev and Nola). Lastly, thank you to Joanna and Orrin: you have been closest to me throughout the writing, and I cannot thank you enough. This book is for you both, with love.

Introduction: Picturing technology

1 Picturing technology today

Picture 'Technology' as a road down which we are travelling, at night. On either side, the lights of some vast metropolis show up as blurs, flashes and streaks. Some argue we are accelerating too fast down this road, and that we may have lost control of our vehicles.[1] Some argue we should accelerate faster, beyond outmoded humanist values holding us back, or to break the economic systems underpinning technology.[2] Some argue we should slow down or turn in the opposite direction.[3] Some argue that this road is in fact a side-track, and that our attention should turn elsewhere.[4] Some argue that becoming 'locked-in' to technologies has left us with a sense that we can only hang on for the ride.[5] Some argue that we should look to the sky, for a God to save us.[6]

Philosophy of technology is the field of contemporary philosophy that aspires to put pictures like this into perspective. It does so by undertaking to systematically investigate the impacts of technologies on one another, and on human and non-human forms of life. While engagements with technologies can be found throughout the history of philosophy, from Plato and Aristotle, to Kant, Rousseau, Marx, Hegel and beyond, 'philosophy of technology' is a relatively young field, and its fortunes as one have been varied. It is commonly held that the term originated with the late-nineteenth-century philosopher Ernst Kapp (Mitcham 1994: 20). In the twentieth century, philosophy of technology can be considered, at best, as a minor subfield of philosophy of science in the analytic philosophical tradition (Bunge 1985). In the continental tradition in philosophy, in contrast, a background concern with technology became so pervasive over the course of the twentieth century as to become a kind of *cliché*. From the phenomenology of Husserl and Heidegger, to the critical theory of Adorno, Arendt and Habermas, to the poststructuralism of Deleuze and Foucault, the basic messages emerging from the continental philosophical tradition on this theme were, it seemed, entirely monotonous: 'Technology' is a road down which we are travelling too fast; it is controlling, rationalizing and deskilling; it colonizes and pollutes the 'lifeworld'; it must be resisted or subverted in favour of more creative forms of life.[7]

Since the late twentieth century, new impetuses in philosophy of technology have emerged.[8] One of the most enduringly influential has been the impetus towards an 'empirical turn'. This emerged from Dutch and North American philosophy of technology in the late 1990s and early 2000s, and, significantly, it tended to take the kind of monotonous continental approach just sketched above as a key target.[9] Against this type of 'classical' continental approach to philosophy of technology, the empirical turn counselled a detailed and pluralistic focus on case studies of technologies in contexts of design, implementation and use (Kroes and Meijers 2000b; Achterhuis 2001; Franssen et al. 2016b). In doing so, it performed some notable services. Conceptually, the empirical turn foregrounded a need to be sceptical of tendencies towards pessimism, determinism and the reification of 'Technology' into some kind of autonomous agent; methodologically, it promoted a more focused and pragmatic approach, against some of the potential pitfalls of more speculative and lyrical ways of thinking about technology evident in the continental tradition.

As this book will argue, however, there is a great deal that is limiting about philosophical 'turns' of this kind. For instance, I will argue that although empirical turn approaches have been right to criticize a continental tendency towards the reification of 'Technology', they have tended to perpetuate the same mistake against the broader continental tradition: by turning a commitment to 'transcendental' method that has been prevalent throughout this tradition since Kant into metaphysical commitment to some form of reified 'Transcendental' realm. Second, I will argue that philosophy of technology in the wake of the empirical turn has often been committed to a sense of what constitutes a 'Technology' that leaves too much to common sense. Third, I will argue that the empirical turn has laid down a problematic precedent for innovation in philosophy of technology, through commitment to a crude picture of 'turning' that seems trivial, but that may in fact have important negative consequences for work in this field.

But why should these points matter beyond a narrow and specialized concern? To hopefully see why, reconsider the picture of the road with which I began. The point is that this is the kind of simplification that is pervasive throughout contemporary thinking about technology: it is, as Wittgenstein famously put it, the kind of picture that can hold us 'captive', because the conditions for it '[lie] in our language' (Wittgenstein 2009: 53).[10] The empirical turn has helped us put aspects of this picture into perspective by calling into question whether 'Technology' can be said to exist as a unified 'capital "T"' phenomenon at all. However, the empirical turn made this move in a limited and narrow way. Faced with the picture of technology as a road, it is as if the empirical turn called for us to stop, get out of our vehicles and inspect the road itself in minute detail, in the hope that the smaller pictures rendered might either dispel the misleading spectre of 'Technology', or that they might scale up into a better approximation of where we are coming from and where we are going.

What this book seeks to offer is a broader and more dynamic sense of what philosophy of technology can and should aspire to today. My premise as to why this matters is that philosophy of technology is a field with a great deal of potential for timely and heterodox thought. As a relatively young and emerging field that is problem-focused and concerned with material culture, philosophy of technology has great potential for the kind of applied, socially engaged and practice-oriented ways of thinking that draw on different traditions and disciplines of thought, and that move beyond entrenched philosophical divisions. It is the contention of this book that, rather than becoming more narrow and specialist in the face of these opportunities, we need to find new ways of mapping the potentials of this field on many different levels of complexity at once: both in order to explore its emergent issues and to chart connections with other fields, including, but not limited to, media theory, software studies, design, engineering and art practice.

What is required for this, I argue, is a more dynamic sense of what constitutes the empirical. It will be the aim of the book as a whole to provide this, but the outlines of its approach can briefly be summarized here. This book will argue:

1. for a conception of the transcendental, not as an otherworldly realm, but as a philosophical method or approach to argument that can be dynamically attuned to the empirical and its conditions. In doing so, it will seek to catalyse what it will position as the immense potential of transcendentally inflected approaches from the continental philosophical tradition for an expansive conception of philosophy of technology today.

2. that our sense of what constitutes a technology might best be examined, not by producing case studies that align with our common sense, but through case studies of 'exceptional technologies' that show up as paradoxical.

3. that we can aspire to a picture of method in philosophy of technology as 'mapping', in contrast to an engrained picture of 'turning'. Rather than turning towards the empirical at the expense of the transcendental, we can aspire to go further in both of these directions at once, towards a sense of the empirical and its conditions that is at once more fine-grained and more wide-ranging.

To explain the concept of 'exceptional technologies' at stake in the second of these points (and in the title of this book) in more detail, let me put things in terms of the picture with which I opened. The argument of this book is that philosophy of technology can and should aspire to help us put our pictures concerning technology into perspective, whether big or small. One way of doing this, I argue, is to expand our sense of the empirical through a lateral move. Framed in terms of the picture with which we began, the point is this: we need to find ways of getting 'off road', to investigate the phenomena that show up as the blurs, flashes and streaks on the margins of such a picture.

'Exceptional technologies' are these 'blurs, flashes and streaks'. That is, they are the phenomena that show up as marginal according to a pervasive common-sense picture of what technology is and where it is taking us, but that in fact act as signs of how this picture connects to much wider issues and diverse forms of life taking place 'off road'. Putting things like this risks seeming captive to the even finer details of a picture; my hope, however, is that the approach developed over the course of this book as a whole will more accurately emerge as a way of 'breaking the frame' (if things have to be put in terms of a picture). To anticipate this, let me state things here in a way that avoids the terms of the picture I have been using so far. The claim advanced in this book is simply this: for any picture of technology, there will be phenomena that show up as marginal and 'exceptional' to it, and these may be just as instructive for helping us make sense of our picture; indeed, in some circumstances, they may be more instructive. In this book, I conceptualize these phenomena as 'exceptional technologies', and I argue that the transcendental approach, as evidenced throughout the continental philosophical tradition, offers a way of being open to them.

To see why this matters, let me risk stressing a specific feature of the picture with which I opened: in a contemporary context where our sense of technologies and their impacts can seem to be accelerating and complexifying greatly, what recommends attention to exceptional technologies is, I think, their capacity to offer a lateral move in favour of instructive critical focus and perspective. The claim advanced in this book, then, is not that exceptional technologies offer something like respite or distraction (as if they were 'laybys' or 'pit stops' for avoiding the intensity of traffic on the road). Instead, the claim is that exceptional technologies might constitute precisely the kinds of intensified 'object lessons' we need today, in a context where our received sense of things can seem to be accelerating towards an increasingly blurred horizon.

The reason for this, as I hope will emerge over the course of this book, is that exceptional technologies are paradoxical, in two senses: they are paradoxical in an etymological sense because they are the kinds of artefacts and practices that run counter to or alongside our received pictures of technology, and they are paradoxical in a logical sense because they are the types of apparently contradictory artefacts and practices that have the capacity to explode the consistency of our settled pictures of technology, and, thereby, to allow us to explore the conditions for these pictures in acute and focused ways.[11] Over the course of this book, I will offer various examples of exceptional technologies, ranging from the case of a blank page in Chapter 2, to the internet in Chapter 3, to case studies of famous merely imagined, failed and impossible technologies in Chapter 4, to a consideration of Foucault's famous treatment of the 'Panopticon' in Chapter 5.

2 Key terms

Since they will each be important to the development of my argument, let me summarize the key terms of this book's title here:

1 *Exceptional Technologies*. This concept is, I hope, the key contribution of this book. By 'exceptional technologies', I mean artefacts and practices that appear as marginal or paradoxical exceptions to a received sense of what empirically constitutes a technology in a given context of design, implementation or use, but that can nevertheless act as important focal points for drawing out and challenging conditions implicated in the received sense. The claim underpinning this book is that, for any given context of technological design, implementation or use, there will be such exceptions. The book offers various examples of exceptional technologies, leading, in Chapter 4, to case studies of a merely imagined technology (Vannevar Bush's 'memex'), a failed technological practice (Francis Galton's 'composite photography') and a technology with an impossible aim (Arthur Ganson's 'Machine with Concrete').[12] The aim of the book is by no means to provide an exhaustive set of exceptional technologies. Instead, it is to establish the beginnings of a project for seeking out further case studies. It is therefore a requirement rather than a disadvantage that the case studies offered in this book should have clear and criticizable limitations: rather than exhausting the concept, my hope is that the limitations of the studies offered in this book might instead be instructive for seeking out different cases of 'exceptional technologies' that can be instructive for other contexts and problems.[13]

2 *Continental Philosophy*. This book links continental philosophy to a sense of the 'transcendental', a theme associated with Kant in the history of philosophy. Significantly, recent developments in both philosophy of technology and continental philosophy have sought to turn away from this theme: just as there has been an 'empirical turn' away from 'Technology' considered as a form of overarching transcendental force in philosophy of technology, so too has there been a putative 'speculative turn' away from Kant's legacy in recent continental philosophy (see Bryant et al. 2011). In Chapters 1 and 5 of this book in particular, I attempt to recuperate a sense of the transcendental from such turns. In Chapter 1, I argue that the

transcendental can be understood in a formal and minimal sense, in terms of a method or approach to argument that involves attention to conditions, and that is not reducible to Kant's doctrine of transcendental idealism. In Chapter 5, I argue that this alternative sense of the transcendental can provide a dynamic and fine-grained account of the empirical that is preferable to that rendered by a logic of 'turning' that the 'empirical' and 'speculative' turns just mentioned share in spite of their otherwise considerable differences. The book's aim in emphasizing a sense of the transcendental is not to suggest that this is the only worthwhile theme in the history of continental philosophy; it is simply to claim that it is an important one, and that, when recognized as such, it provides a way of foregrounding overlooked rigours, coherences and cross-disciplinary trajectories in the continental tradition.[14]

3 *Philosophy of Technology.* This book considers the empirical turn as an entry point for making sense of recent developments in philosophy of technology. This raises three big concerns: first, it might appear to fixate on a localized event in a highly specialized field; second, it might appear to fixate on what is by now a dated event in this field (Franssen et al. 2016b); and third, it might appear contrived to make a straw man of 'empirical turn' philosophers for rhetorical purposes. In an attempt to dispel these worries, let me note three main reasons why this book focuses on the empirical turn. The position advanced in this book is: first, that the notion of an 'empirical turn' is not adequate to describe the richness of what is at stake in philosophical considerations of technology, including putatively 'empirical turn' ones; second, that the force of the empirical turn is in fact normative, and that it sets a misleading norm for the type of work that philosophers of technology can and should do, because it involves turning away from other important aspects of this field and because it may limit the potential for cross-disciplinary work; and third, that the empirical turn takes its normative warrant mostly from unclarified common-sense presuppositions on what constitutes a 'technology', and on why and how we should be turned towards such entities.[15] These points do not construe the 'empirical turn' that took place in the late 1990s and early 2000s as synonymous with philosophy of technology today. On the contrary, they presuppose that the concept of an empirical turn is inadequate for describing what has gone on in philosophy of technology to date, and that it is divisive as a norm for what can and should go on in the field in the future.

3 Structure and limits

The book is structured into five main chapters. Chapter 1 outlines a sense of the transcendental as a method or approach to argument. I begin by foregrounding Catherine Malabou's *Avant demain* as an example of a recent reflection on the status of the transcendental from within the continental tradition. The chapter then seeks to show how a sense of the transcendental is not reducible to the terms of Kant's 'transcendental idealism'. After this, I consider the epistemological reception of transcendental arguments in analytic philosophy, followed by a consideration of the ontological sense of the transcendental in the continental tradition since Heidegger. I conclude this chapter by outlining some potential benefits that the sense of the transcendental developed in it might bring to philosophy of technology in the wake of the empirical turn.

Chapter 2 considers a problem of relevance: what is to stop the sense of the transcendental developed in this book from spilling into an infinite regress of conditions? The chapter responds with a case study in favour of a sense of Edmund Husserl's practice of 'imaginative variation' as historically and materially situated. The argument is that imaginative variation can, on this reading, be viewed as a practical enactment of a sense of the transcendental that blocks its theoretical tendencies towards infinite regress. I focus on a case study of the 'blank page' to focus this. I begin by considering the conditions under which this example is imagined in Husserl's *Ideas* and *The Crisis of European Sciences and Transcendental Phenomenology*. After this, I consider how these conditions have varied today, with a focus on ways in which our contemporary sense of the 'blank page' has been figuratively extended through innovations in computing. The chapter concludes by positioning an expanded sense of imaginative variation as a potential 'continental' complement to post-empirical turn attempts to extend the scope of philosophy of technology that take their inspiration from the analytic philosophical tradition.

Chapter 3 seeks to show that approaches drawing on a sense of the transcendental are already implicit across diverse contemporary philosophical considerations of technology. It does so by focusing on approaches that emphasize the importance of embodiment conditions. The chapter begins with a critical reading of Hubert Dreyfus's *On the Internet*. I argue that Dreyfus demonstrates a sense of the transcendental that is limited, but that points towards the sense in which the internet can be considered as an 'exceptional technology'. I then seek to take Dreyfus's sense of the transcendental further through a focus on the approaches of two well-known figures in recent media theory: N. Katherine Hayles and Mark B. N. Hansen. The chapter concludes by considering how the '4e' approach to cognitive science – as evidenced in the work of thinkers including Mark Rowlands, Shaun Gallagher and Andy Clark – can be viewed as taking a sense of the transcendental further still, in a sense that is consistent with philosophical naturalism (Rowlands 2010).

Chapter 4 can, I hope, be viewed as the centrepiece of this book. While the chapters leading up to it seek to outline a sense of the transcendental as a way of being open to exceptional technologies, this chapter aims to provide three focused case studies: of a merely imagined technology (Vannevar Bush's 'memex'), a failed practice (Francis Galton's 'composite photography') and a technology with an impossible aim (Arthur Ganson's 'Machine with Concrete'). As intimated above, the aim of this chapter is not to provide a definitive set of exceptional technologies. Instead, it is to show how exceptional technologies can be implicated in actual processes of technological design, implementation and use, and to offer a non-exhaustive entry point for the concept. The chapter also aims to show how we might draw on a number of different disciplinary perspectives to inform work in philosophy of technology, and to show how the concept of exceptional technologies may be of cross-disciplinary interest. This chapter can, I hope, be read as a stand-alone piece, as, I hope, can the case studies included in it. As I emphasize in concluding this chapter, its three case studies should each have recognizable and criticizable limits. This is because they are intended as entry points for seeking out further case studies that might be instructive for different problems and contexts.

Chapter 5 considers potential objections to the senses of 'continental philosophy' and 'philosophy of technology' used throughout this book. It might seem that my sense of continental philosophy does not pay enough attention to a recent 'speculative turn' in this tradition. Conversely, my sense of philosophy of technology may seem to pay too much attention to the 'empirical turn'. This chapter seeks to dispel these objections by focusing on a shared picture of 'turning' that can be viewed as underpinning the speculative and empirical turns alike: for every turn towards, according to this picture, attention must turn away from something else, and this, I argue, involves a picture of method that is too crudely first person, oppositional and progressivist to help us engage complexities that go into shaping philosophy of technology as a field. An alternative picture of method, I argue, can be rendered with a picture of 'mapping'. However, just as with the picture of 'turning', we have to be wary of becoming captive to this alternative picture. To draw these issues out, the chapter develops a reading of Foucault's work on Bentham's 'Panopticon' in *Discipline and Punish*, which I position as an exemplary attempt to 'map' an exceptional technology, on multiple levels of complexity at once.

I conclude the book by distinguishing its approach from what might be called technological 'exceptionalism'. Building on the conclusion of Chapter 5, I argue that exceptional technologies should not be viewed as a class of entities to be privileged across all possible contexts. Instead, I argue that they are valuable as paradoxical object lessons for considering what constitutes an object as 'technological' in a given context.

To close this introduction, let me briefly note some of the limitations of this book. There are a number of thinkers who merit more attention than they receive. In particular, the work of Andrew Feenberg, Gilbert Simondon, Bernard Stiegler and thinkers in the 'media archaeology' movement has perhaps not been engaged with in sufficient depth. Since the aim of this book is not to give a comprehensive survey of continentally inspired approaches to philosophy of technology, but instead to articulate a philosophy of technology that draws on the continental tradition, however, I hope that I might be granted these omissions. Given more time and space, these thinkers would have figured more prominently; the fact that they do not simply means that there are clear trajectories for future work.

CHAPTER ONE

A sense of the transcendental

The aim of this chapter is to show why and how a renewed sense of the transcendental as an approach to argument or method might make sense for philosophy of technology today. In Part 1, I engage with recent remarks on the status of transcendental reasoning in contemporary philosophy from Catherine Malabou. I cite Malabou here for two main reasons: first, to foreground rhetorical gestures that might, on an uncharitable and reductive reading, be viewed as symptomatic of some of continental philosophy's worst tendencies; second, and more importantly, to endorse remarks she makes on the significance of Kant and a sense of the transcendental for the continental tradition in philosophy. This part concludes by contextualizing Malabou's approach to two challenges facing continental philosophy today: a 'speculative turn' emerging from within the continental tradition itself, and a materialist challenge to the presupposed 'irreducibility' of the transcendental, as exemplified for Malabou by the reductionist programme of contemporary neuroscience.

Part 2 outlines the history behind the sense of the transcendental I am investigating. In contrast to a tendency towards reification of this term, I argue for a sense of the transcendental as an approach to argument or method. I begin by showing how this sense is not reducible to Kant's doctrine of 'transcendental idealism'. After this, I chart how it has played out in subsequent approaches, drawing first on the sense of transcendental arguments prevalent in the epistemological tradition in analytic philosophy, then on the ontological sense of the transcendental forwarded by Martin Heidegger.

Part 3 seeks to articulate this book's sense of the transcendental in as clear a way as possible. It does so by radicalizing Jeff Malpas's reading of the hallmark of the transcendental as a 'circularity of structure' (Malpas 1997: 4). By comparing Kantian and Heideggerian approaches to the transcendental, and by working through the stages of Malpas's reading,

I argue for a minimal formal sense: given 'X', a transcendental approach is one that enquires into *a priori* conditions for X. This part closes by considering the ways in which the apparent triviality of this minimal sense in fact gives dynamic and complex range. This, I argue, is because reflection on each of its key terms ('given', 'X', 'inquire', '*a priori*', 'conditions') can lead to a complex, nuanced and evolving sense of the empirical. As such, my operative sense of the transcendental is not merely circular, but recursive.

Part 4 seeks to directly show why and how this approach might make sense for philosophy of technology today. Since the empirical turn, 'transcendental' tends to be used in philosophy of technology as a synonym for the bad essentializing tendencies of so-called 'classical' philosophers of technology such as Jacques Ellul, Hans Jonas, Karl Jaspers and, most notoriously, Heidegger (Achterhuis 2001; Brey 2010; Verbeek 2005: 1–12). This part develops three main criticisms of this way of approaching philosophy of technology in the wake of the empirical turn. First, I argue that it tends to repeat a fallacy of reification that it diagnoses in 'classical' approaches. Second, I argue that it tends towards problematic common-sense presuppositions on what constitutes a 'Technology', to the detriment of 'exceptional technologies'. Third, I argue that it has set a problematic precedent where the key picture for innovation in philosophy of technology is one of 'turning'.[1]

1 Malabou's sense

Consider these remarks from Catherine Malabou's 2014 book *Avant demain*:

> The transcendental: to save or destroy it, to transform or make it derivative, to temporalize it or to break with it? As we have seen, it is most often the case that conservation and abandonment coincide. ... Kant is not any old philosopher. ... He is the guarantor of the identity of continental or 'European' philosophy. This latter, the visibility and institutional power of which never cease to shrink everywhere in the world, doesn't it owe its specificity precisely to the claim that something like the transcendental exists? Something which Kant presents as the form of thought, and which only has theoretical and practical reality for thought as such? Incontestably, adherence or opposition to the transcendental marks better than all the other criteria the fracture between the continental and analytic traditions [in philosophy], that is, between two comprehensions of rationality. (2014: 222–3. My translation)

Malabou's key claim here is that continental philosophy owes its specificity to the claim 'that something like the transcendental exists'.[2] Before unpacking this point, however, let me foreground several features of *how* she makes it.

Malabou frames her claim in terms of a pessimistic observation about the destiny of continental philosophy and in terms of a rhetorical question. The immediate problem is that these might appear, on an uncharitable reading, to be precisely the kinds of generically 'continental' philosophical moves that bring in too much obfuscation and rhetoric at the price of clarity and precision. When we set Malabou's remarks in the context of those immediately following them, moreover, this problem might seem only to get worse:

> The problem is that adherence to the transcendental ... far from being univocal, is often already in itself an opposition to the transcendental. Every post-Kantian attempt to safeguard the transcendental always reveals itself, in one way or another, to be an attack against the transcendental itself. In effect, the inheritors of Kant are always split ... between two visions of the transcendental: the one *hyper*-, the other *hypo*-normative. (2014: 223, My translation, Original emphasis)

How might this passage appear to an uncharitable reader of continental philosophy? First, its remarks concerning the unity of opposites may seem too enamoured with the legacy of deconstruction, with dialectics or simply with a taste for paradox and the counterintuitive. Second, its contextualization in terms of 'post-Kantian' philosophy may seem irrelevantly historical or 'genealogical', or too ambiguous or ambitious.[3] Is Malabou using 'post-Kantian' here to classify the German idealism of Hegel, Fichte and Schelling; the phenomenology of Husserl and Heidegger; the French poststructuralism of Derrida and Deleuze; or all of the above? More uncharitably still, is she simply tending towards jargon? What, for example, is meant by her terms 'hyper-normative' and 'hypo-normative'?

The reason for raising the spectre of this uncharitable reading here is not to indulge it, but to exorcize it.[4] With Malabou, this chapter will take the claim that continental philosophy is intimately tied to a sense of the transcendental to be accurate and important. In doing so, it will deviate from some of the particulars of her approach. To get to a stage where such deviations are possible, however, we have to be prepared to raise and dispel certain preconceptions concerning the limits and purposes of 'continental' philosophy that are prevalent throughout contemporary philosophical culture, as Malabou herself recognizes.

To develop this point, consider the following passage where Malabou moves to deflate the apparent jargon of the terms 'hyper-normative' and 'hypo-normative':

> According to [the hyper-normative view], the transcendental represents a sort of censure which absolutely forbids every mixture with experience, and, consequently, every becoming and transformation of logical forms. ... Following the logic of [the hypo-normative view] the

> transcendental is certainly a constraint – of form and of structure – but this constraint, paradoxically, is a synonym for liberty. It is supposed to guarantee, in effect, the autonomy of thought with regard to every determinism or reductionism. The transcendental then defines itself as *irreducibility*, pure symbolic latitude. (2014: 223, My translation, Original emphasis)

Another way to describe the 'hyper-normative' approach, on this account, would be 'conservatively Kantian' (Malabou 2014: 67, 78). This approach takes 'transcendental' to refer to conditions for the possibility of thought and experience that are *a priori* in Kant's sense: that is, as having to do with necessary and universal structures of thought (such as Kant's famous 'categories' of the understanding) that allow the subject to make sense of experience, but that are not generated by experience, and that are preserved from 'all risk of empirical contamination' (Malabou 2014: 67–8).

Another way to describe 'hypo-normative', on the other hand, would be as more or less 'liberally' or 'critically' Kantian. *Avant demain* aligns this approach with thinkers including Foucault, Ricoeur and Derrida (Malabou 2014: 171–89, 222). The tendency of these thinkers, Malabou holds, is to emphasize the spirit of Kant's approach against the more conservative tendency to interpret the *letter* of Kant's doctrine of 'transcendental idealism'. Put very crudely, whereas hyper-normative approaches are interested in what Kant himself meant by 'transcendental', hypo-normative approaches are more interested in how a sense of the transcendental can be taken up for engaging philosophical problems that are not necessarily or obviously 'Kantian'.

But what is there in all of this that makes 'the transcendental' anything more than a historical matter? As Malabou notes:

> Today, this constant ambiguity [between 'hyper' and 'hypo' normative approaches], indeed this contradiction in the vision of the transcendental – policing or permissive – appears in full view. It is no longer possible to hide it, in the name of the presumptively unsurpassable character of Kantianism. On the one hand, this is because the censure is lifted: it is possible to break the lock of the transcendental. ... On the other hand ... materialism, once again, is seeking to make itself heard. The frontier between 'thought' and the cerebral is becoming more and more difficult to define, for example. To maintain the existence of the transcendental at all costs in the name of antireductionism coincides more often than not with a reactive position. (2014: 224–5, My translation)

As touched on above, Kant's transcendental idealism famously presupposes the objects of experience to be 'appearances' constituted by *a priori* forms of thought and experience found in the subject, and not in 'things in themselves'. What has been 'presumptively unsurpassable' for continental philosophy since Kant, on this reading, is a form of 'correlationism' that centres the transcendental (the form and structure of thought and experience) and the empirical (the actual and possible objects of experience) on the mediating figure of subjective consciousness. As Malabou notes in the above passage, a recent attempt to break this 'lock' has emerged in the continental tradition in the form of a 'speculative turn' that seeks to determine features of objects that transcend these putatively subject-centred limits (see Meillassoux 2006: 155–78).[5] Independently, but complementarily, Malabou notes that developments in contemporary neuroscience and biology emerge as a practical challenge for correlationism; this is because what they suggest is that the correlation between subject and object can be reduced to a purely naturalistic explanation, without any elusive 'transcendental' remainder in the subject.

Caught in this pincer movement, one trajectory for a sense of the transcendental would, as Malabou notes at the end of the above passage, be a retreat into dogmatism. Interestingly, however, a feature of Malabou's account suggests that other (non-dogmatic) senses of the transcendental may still be possible. Note Malabou's qualification that attempts to maintain the existence of the transcendental lead to a dogmatic and reactive position 'more often than not'. Either this is a throwaway figure of speech or it is the conclusion of an argument that Malabou does not explicitly present: If it is the former, then it would have to be disqualified as a proper injunction against exploring the possibility of a non-dogmatic sense of the transcendental; if it is the latter, then it is arrived at through inductive reasoning, and this does not logically prohibit the possibility of such an approach either.

Instead of reading Malabou's remarks as an injunction against the possibility of non-dogmatic approaches to the transcendental, my aim in this chapter is to read them as a challenge to articulate such an approach.[6] In doing so, I will not follow Malabou's own emphasis on neuroscience and biology; instead, I will shift the emphasis to philosophy of technology.

To see the rationale for this shift, reconsider Malabou's above remarks that 'materialism, once again, is seeking to make itself heard', and that 'the frontier between "thought" and the cerebral is becoming more and more difficult to define' (2014: 224). Since 1996s *The Future of Hegel*, Malabou has been providing a unique and powerful take on developments in neuroscience and biology that draws on the continental philosophical tradition, with an emphasis on the theme of 'plasticity' (Malabou 2004, 2005, 2009; James 2012: 83–109). This background is at work in the passage just cited, and underpins her example of the frontier between 'thought' and the cerebral. If Malabou's claim that 'materialism ... is seeking to make itself heard' is accurate, however, then there is something her example understates: the multitude of other ways this 'announcement' must be occurring, across different fields.

What emerges here is an initial sense as to why philosophy of technology emerges as a potentially significant field for exploring a sense of the 'transcendental': it is a field where the possibility and status of the correlation between thought and material culture has always been at stake in complex and plastic ways, and where materialism's announcement of shifting frontiers, if accurately diagnosed by Malabou, must be 'making itself heard'. The question that now arises is this: is there a non-dogmatic sense of the transcendental that can help us make sense of such a field?

2 Expanding sense

Recall the passage I cited from Malabou at the beginning of Part 1. The transcendental emerged from it as a very special kind of 'something', open to all kinds of actions being performed on it, and intimately connected to Kant and the continental tradition in philosophy.

This part will argue that it is partly misleading to frame the transcendental in this way. This is because tendencies to reify the transcendental as an allusive 'something' are, I will argue, apt to cut it adrift as a metaphysical realm opposed to, and out of step with, the empirical. In contrast, I will argue that a more philosophically acute move consists in recognizing 'the transcendental' as a dynamic form of philosophical approach to argument or method. Put even more simply, I will argue that the word 'transcendental' makes better sense for philosophy as an adjective than as a noun. Rather than constituting a substantive entity or set of entities existing out of touch with the empirical (as nouns like 'the transcendental' and 'the transcendental realm' suggest), I will argue that 'transcendental' better describes a critical and creative approach to doing philosophy that can be immanently and dynamically responsive to an evolving sense of the empirical, understood in the Kantian sense of the actual and possible objects of experience.

An important first move for articulating this sense of the transcendental consists in showing that it is irreducible to Kant's doctrine of 'transcendental idealism'. As Sebastian Gardner writes:

> Transcendental idealism has ... to state the obvious, a complex structure. Kant proceeds on distinguishable levels, on the one hand focusing on and interpreting the very concept of objecthood, and at another level advancing theses correlating specific formal features of objects with specific features of our mode of cognition. We may refer ... to the former as Kant's transcendental turn, and to the latter as Kant's transcendental idealism. (2015: 7)

Although I will demur from Gardner's use of the term 'turn' throughout this chapter (and, indeed, throughout this book), the important point here is that a commitment to transcendental idealism does not have to follow from a sense of the transcendental.[7] 'Transcendental idealism', to recap, is Kant's metaphysical doctrine that objects as we experience them are constructed by forms of thought common to all cognizing subjects, and that, as cognizing subjects, we can only know objects as they appear under these conditions, never as they are 'in themselves'. In Kant's own terms, it is the doctrine that 'we can know *a priori* of things only what we ourselves put into them', through what he calls 'synthetic *a priori*' judgements (Kant 2000: B xviii). In contrast to this, what Gardner points us towards is a sense of the transcendental that is irreducible to this, and that shows up as a methodological innovation in philosophy.

The core issue at stake for this sense of the transcendental concerns 'focusing on and interpreting the very concept of objecthood'. Transcendental approaches are therefore concerned with the question, 'How are objects possible?' They can ask this question either of objects in general, or of particular sets of objects, and their hallmark in addressing it consists in deploying a regressive form of argument that moves from objects to their necessary conditions of possibility. In literature on transcendental arguments in the epistemological tradition, definition of this approach is usually the first issue to be confronted. As Paul Franks puts it:

> What are transcendental arguments? ... If philosophy begins in wonder, transcendental philosophy begins with wonder that takes the form of a characteristic question: 'How is X possible?', construed as 'What are the necessary conditions for the possibility of X?' ... The conditions of possibility with which transcendental philosophy concerns itself may be variously construed as causes, reasons, or as circumstances without which (in fact or in logic) something could not arise, or be, or endure. (Franks 1999: 113–15)

Again, no explicitly Kantian 'transcendentally idealist' metaphysical commitments seem to be involved here. Instead, a form of argument is highlighted that exceeds the scope of explicitly Kantian philosophy in several ways. The most important of these concerns the broadening of the scope of transcendental arguments to incorporate 'circumstances' and 'factual' conditions of possibility, in contrast to Kant's emphasis on purified 'transcendental logic'. In seeking out signs that Franks takes a separable sense of the transcendental to be possible, however, his opening question seems sufficient: by asking 'what is a transcendental argument?', and not 'what is a transcendental *idealist* argument?', he appears to presuppose that the former is irreducible to the latter.

The feasibility of such a form of argument is often presupposed throughout the literature.[8] Take, for instance, the following from Charles Taylor:

> The arguments I want to call 'transcendental' start from some feature of our experience which they claim to be indubitable and beyond cavil. They then move to a stronger conclusion, one concerning the nature of the subject or the subject's position in the world. They make this move by

a regressive argument, to the effect that the stronger conclusion must be so if the indubitable fact about experience is to be possible (and being so, it must be possible). (1995: 20)

What again is at stake here is a form of argument that can be employed by philosophical approaches with different epistemological and metaphysical commitments to those of Kant. What is particularly worth noting in this respect is that the essay from which this passage is taken is concerned with Merleau-Ponty. When Taylor discusses features of experience that are claimed to be 'indubitable and beyond cavil', then, he is more concerned with the conditions of embodied experience than 'synthetic *a priori*' judgements (Taylor 1995: 23–4), and what he has in mind on 'the nature of the subject or the subject's position in the world' has more to do with the experience of being dynamically oriented in space, rather than Kant's more abstract notion of 'categories' of the understanding.[9]

Not all approaches are sanguine on the merits of reading the transcendental approach as irreducible to Kantian commitments, however. Jaakko Hintikka, writing in the wake of Strawson's famous work on transcendental arguments in *Individuals* (1959), argued for a much more limited sense that situates transcendental arguments squarely within the post-Cartesian 'way of ideas' tradition in epistemology (Hintikka 1972; Beiser 2002). On this interpretation, it is vacuous or, worse, deeply misleading to speak of a sense of the transcendental that is irreducible to Kant's metaphysical commitments, because this is 'beside the point' of what Kant himself sought to achieve. Specifically, Hintikka holds that Kant took transcendental arguments to be attempts to refute the sceptic on such issues as the existence of the external world and other minds (1972: 275–7).[10] For Hintikka, the problem with any sense of the transcendental that seems irreducible to this is that it stands merely to obscure these epistemological issues, as well as Kant's contributions to identifying and addressing them.

Other thinkers who adopt a restrictive approach include Robert Stern (1999) and Barry Stroud (1968, 1999). In his famous 1968 article, 'Transcendental Arguments', Stroud worked within the epistemological paradigm to argue against the validity of strong transcendental arguments that require transcendental idealist premises (Stroud 1968). Subsequently, he moderated this by highlighting the potential benefits of more 'modest' transcendental arguments as strategies for drawing out and problematizing shared/foundational beliefs in epistemic communities (Stroud 1999: 163; Hookaway 1999; Stern 1999: 2). While Stroud is favourable to a modest interpretation of the sense of the transcendental highlighted above by Gardner, then, he is hostile to any form of 'strong' approach that would attempt to infer how things must be in the world (metaphysically) from how they are believed to be in a knower or community of knowers (epistemologically). As a corollary of this, Stroud goes further than Hintikka by claiming that the notion of a strong transcendental turn is not merely 'beside the point', but that it might be immodestly enthralled to a naïve form of transcendental idealism. As he puts it in a passage that is worth citing at length:

> What I question are the efforts of those who retain what looks like the generally Kantian strategy while dropping the transcendental idealism that was supposed to explain how the whole enterprise was possible and could yield positive results. ... Put in the most schematic terms, what is problematic is that the conclusions of the most ambitious transcendental arguments without transcendental idealism are apparently meant to state how things are – that there are enduring objects, for example, or that events are related causally, or that there are persons with thoughts and feelings, and so on – and in a [transcendental and not simply psychological and subjective] way that in itself says nothing about anyone's thinking or believing that things are those ways. But such conclusions about the world are to be reached transcendentally by *a priori* reflection on the conditions of our thinking and experiencing the things we do. That appears to mean that transcendental reflection starts from statements like 'We think or experience in such-and-such ways' or 'We believe that things are so-and-so'. We start with what we can call psychological premises – statements whose main verb is a psychological verb like 'think' or 'believe' – and somehow reach non-psychological conclusions which say simply how things are, not that people think things are a certain way. (1999: 160)

The key point here is that, for Stroud, approaches that emphasize a strong sense of the transcendental that seeks to go beyond Kant in fact emerge as more confused versions of 'what Kant sought' (1999: 160). What such approaches actually stand to pursue, on this reading, is a form of transcendental idealism that is ungrounded, vague and much more metaphysically naïve and psychologically 'subjectivist' than that of Kant.

But what happens if we do not presuppose 'how things are' to be exhaustively covered by the typical 'way of ideas' epistemological problematics Stroud cites in the above passage ('that there are enduring objects ..., or that events are related causally, or that there are persons with thoughts and feelings, and so on')? And, relatedly, what happens if we question Stroud's notion of the subject of such an inquiry (i.e. if we question the 'we' of his 'we believe that things are so-and-so')?

What I am alluding to here is a sense of the transcendental that would exceed the scope of the epistemological paradigm, and that would be more specifically 'continental' in character. In investigating this, my aim is not to undo important work in the epistemological paradigm; instead, I am simply attempting to see if issues pertaining to our sense of the transcendental might have purchase beyond its purview. With this in mind, we are in fact returning to two key issues identified in relation to Malabou in Part 1 of this chapter. First, what is the specificity of 'continental' philosophy's approach to the transcendental? Second, must 'transcendental' philosophy always involve a subject–object correlation?

There are many possible starting points for investigating a specifically 'continental' approach to the transcendental. With Franks, one might begin with the post-Kantian German idealism of Reinhold, Hegel, Fichte or Schelling (1999: 111–45). With Zahavi or Moran, one might begin with Husserl's 'transcendental phenomenology' (Zahavi 2015: 228–43; Moran 2007: 135–51). With Bennington, one might start with the 'quasi-transcendentalism' of Derrida (2002). With Sauvagnargues and Bryant, one might start with the 'transcendental empiricism' of Deleuze (Sauvagnargues 2008; Bryant 2008). With Malabou, one might start with the 'genealogy' or 'archaeology' of Foucault (2014: 171–89). Perhaps the best starting point, however, is with Heidegger.

There are three main reasons for this. First, Heidegger is an influential historical bridge between German idealism, Husserlian phenomenology and, further, the approaches of thinkers like Derrida, Deleuze and Foucault. Second, Heidegger's approach comprises a much more controversially 'ontological' touchstone than either German idealism or Husserlian phenomenology when viewed through the lens of the epistemological tradition to which thinkers such as Hintikka, Stern and Stroud belong, and this means that it can help draw out key areas of difference in acute ways. Third, Heidegger is, as I will seek to draw out in greater detail in the final part of this chapter, a controversial but influential thinker in philosophy of technology: foregrounding his approach to the transcendental here, then, will allow me to prepare some context for that consideration.

To get started on these issues, consider the following remarks from Crowell and Malpas:

> Perhaps the most well-known feature of Heidegger's Kant interpretation is his rejection of [the] 'epistemological' reading of Kant; instead he favours the claim that Kant's enterprise was really an 'ontological' one. ... This widening of the scope of the transcendental question stemmed from what Husserl's transcendental phenomenology had already accomplished, namely, a 'break with the way of ideas', that is, a break with an understanding of intentionality as something that is mediated by mental 'representations'. To understand transcendental philosophy essentially as an answer to a certain kind of scepticism (that is, as primarily an epistemological enterprise) is to remain within the Cartesian framework in which alone such a problem can arise. Heidegger's reading of Kant makes explicit the tension within Kant himself between a residual Cartesianism and a new paradigm, in which mind is always in the world and subject and object cannot be thought as separate. (2007: 3)

This passage condenses all the key issues discussed up to now over the course of this part. First, it makes it clear that Heidegger takes up a radical sense of the transcendental that construes it as irreducible, not merely to the concerns of Kant's transcendental idealism, but to the epistemological paradigm in modern philosophy per se. Second, with reference to the two questions raised above in connection to Malabou, the passage offers answers: on the specificity of 'continental' philosophy's approach to the transcendental, it intimates that a widening of scope beyond epistemology is what is definitive (in Heidegger's case, this takes on the character of 'ontological' inquiry); on whether a 'correlation' is always implied in transcendental philosophy, it answers, in the last sentence, explicitly in the positive, and in a way that marks Heidegger's approach out as an archetypal form of what Quentin Meillassoux has called 'strong correlationism' (2006: 31–5).

Here, reflecting back on *Being and Time*, is how Heidegger puts the first matter concerning what it means to move from the epistemological to the ontological:

> If we radicalise the Kantian problem of ontological knowledge in the sense that we do not *limit* this problem to the ontological foundation of the positive *sciences* and if we do not take this problem as a problem of *judgement* but as the radical and fundamental question concerning the possibility of *understanding being in general*, then we shall arrive at the philosophically fundamental problematic of *Being and Time*. (1997: 289. Original emphasis)

As this passage implies, the key Heideggerian move in the search for an expanded sense of the transcendental is to assert that our (i.e. '*Dasein's*') understanding of 'being-in-the-world' is not founded on explicit judgements articulated in the form of propositions. Instead, Heidegger holds that implicit non-propositional forms of understanding are required as transcendental conditions for the possibility of such judgements.

There are, for Heidegger, two important respects in which the Kantian approach to the transcendental, along with that of subsequent epistemology, is insufficiently expansive or 'ontological' on this account. First, to the extent that such approaches presuppose the findings of our best current positive sciences to be bodies of explicitly formulated and testable propositions, they do so to the exclusion of more implicit modes of inquiry, such as 'mood' and, crucially, the structure that Heidegger calls 'care' (Heidegger 2005: 225–73). Second, to the extent that such approaches presuppose explicit judgements framed in the form of propositions to be the fundamental items of knowledge and experience, they do so to the exclusion of prior non-propositional processes of 'ready-to-hand' pragmatic 'doing' and 'coping', with entities as 'equipment' (Heidegger 2005; Dreyfus 1991). As Mark Okrent puts it:

> Th[e] ability to intend things as belonging to or adhering to equipmental types, as ready-to-hand, provides the base step for all of Heidegger's

> transcendental arguments. ... Just as Kant ... argues that the ability to conceptually cognise objects in judgements is necessary for the ability to intend a single unified world of possible experience or empirical knowledge, Heidegger argues that the ability to intend entities as ready-to-hand is necessary for a variety of other kinds of intentional comportments. (Okrent 2007: 161; see also Lafont 2007; Blattner 2007)

There is, I think, something both reassuring and deceptive about the guiding thread Okrent offers here. It is reassuring because it locates 'the base step for all of Heidegger's transcendental arguments' in one of the most famous features of his philosophy: the position advanced in *Being and Time* in favour of the ontological priority of the 'ready-to-hand' over 'present-at-hand' theorizing about entities as 'objects'. As is consistent with the passage from Heidegger cited above, Okrent's point is that this prioritization is itself a form of 'transcendental argument' insofar as it takes the ready-to-hand to be the more 'primordial' condition for the present-at-hand.

What is deceptive is that this understates problems involved in justifying such a claim. The key problem is how the prioritization Heidegger seeks to effect could be called an 'argument' at all. Since Heidegger seeks to step outside the epistemological paradigm within which arguments can be said to be cogent or valid, his approach either seems to fall into the contradiction of requiring a condition it disavows, or, alternatively, to be based on a bare ontological assertion concerning the way things are. Given Heidegger's indebtedness to phenomenological description as a starting point for philosophy (Heidegger 2005: 51–63), such an alternative need not be fatal, but whether or not it is credible will turn on the type of phenomenology the reader favours. If the reader is simply not convinced by Heidegger's descriptions, then the problem is that what will really stand out from his putatively 'ontological' approach are its subjectivist idiosyncrasies (e.g. predilections for hammers and workshops, for etymology, not to mention deeply problematic issues concerning his politics).[11]

What we encounter here is the force of the second key question raised above, concerning the sense in which Heidegger's approach is a form of 'strong correlationism'. Correlationism, to recap, is the view that it is meaningless to talk about objects outside the correlation between an experiencing subject and objects as they appear for it. On Meillassoux's account, Heidegger's approach to the transcendental is, in contrast to the 'weak correlationism' of Kant's transcendental idealism, a form of 'strong correlationism'. This is because, whereas Kant holds objects 'in themselves' to be at least thinkable (if not knowable or experienceable), Heidegger, on Meillassoux's account, takes all such talk of objects outside of the correlation between *Dasein* and entities as they 'show themselves' to be meaningless (Meillassoux 2006: 31–6; Sparrow 2014: 88–90; Braver 2013).

Insofar as Kantian things in themselves act as thinkable constraints on what is objectively knowable (and, crucially, as developed in the *Critique of*

Practical Reason (Kant 1996), on what is objectively appropriate in terms of ethical action), they act as barriers to subjectivist excesses. As a rule, then, we can say that the stronger a form of correlationism on Meillassoux's account, the more susceptible it will be to subjectivist excesses of the type intimated by Stroud in the passage quoted at length above: Heidegger's approach to the transcendental emerges as a paradigmatic example of 'strong correlationism' on Meillassoux's account, and it therefore emerges as very susceptible indeed to the kind of subjectivist excesses highlighted by Stroud.

3 Expanding further: From minimal to maximal sense

It may seem we have reached the bounds of the transcendental in shifting from Kant to Heidegger. But what if there is a different way to make sense of this theme? To pursue this, this part will consider the drift of the argument of Jeff Malpas's essay 'The Transcendental Circle'. At the beginning of this piece, Malpas writes:

> ['Transcendental argument' means] a form of reasoning that proceeds from the fact of experience to the necessary conditions on which the possibility of such experience rests. (1997: 3)

Malpas's tactic here echoes the approaches of Franks and Taylor discussed in the last part, and is formal enough to be consistent with both the epistemological approaches of Hintikka, Stern and Stroud, and with Heidegger's 'ontological' approach. Witness, however, how he moves to take the inquiry further:

> Of course, it is often pointed out that Kant himself took the transcendental to refer back to the constituting power of transcendental subjectivity as the ground for experience and knowledge, and this may in turn be taken to suggest that transcendental arguments should be characterised by reference to the idea of the self-constituting subject rather than by reference to any circularity of structure. It is certainly clear that Kant did take the transcendental ground of experience to lie in the subject and that he saw transcendental philosophy as characterised by reference to the self-constitution of subjectivity. ... This way of understanding transcendental philosophy is nevertheless not independent of the idea of the transcendental as concerned with the question of the possibility of experience and with the attempt to ground experience (or knowledge) by reference to experience itself. ... To treat transcendental philosophy as always leading back to the transcendental subject would be to identify

Kant's particular view of the conclusions that must be reached by such argument with what is essential to the argument as such. (1997: 4)

What appears essential to transcendental inquiry according to the epistemological paradigm, to recap, is a form of argument that, by focusing on *a priori* subjective conditions for the possibility of knowledge, seeks to refute scepticism on such classic issues as whether 'there are enduring objects ..., or that events are related causally, or that there are persons with thoughts and feelings' (Stroud 1999: 160). As Malpas notes, however, such a focus is a contingent feature of transcendental inquiry:

> It is often taken for granted that transcendental arguments are characteristically arguments designed to refute the epistemological sceptic. ... Transcendental arguments may well have anti-sceptical consequences, but to take such arguments as characterised by their anti-sceptical consequences would once again be to take those arguments as characterised by their conclusions rather than their particular structure. (1997: 5)

Given the possibilities for making sense of the theme of the transcendental discussed in the previous part, this shift away from the concerns of the epistemological paradigm may appear destined to commit Malpas to Heidegger's 'ontological' paradigm. While it is certainly true that Malpas's overall approach is greatly indebted to Heidegger, however, it should be emphasized that a key feature of his essay turns on an overlooked aspect of Kant's approach: the fact that Kant characterized his own approach as 'ontological':

> The *Critique of Pure Reason* is by its author's own account a work, not of epistemology, but of ontology. Indeed, Kant claimed that ... 'ontology ... will be called transcendental philosophy because it contains the conditions and elements of our *a priori* knowledge'. (Malpas 1997: 1)

The passage from Kant that Malpas cites here is drawn from the relatively obscure essay 'What Real Progress Has Metaphysics Made in Germany since the Time of Leibniz and Wolff?' This piece was begun in 1793 but was never published in Kant's lifetime.[12] It might therefore be tempting to try to write off Kant's comments as idiosyncratic in relation to the dominant epistemological interpretation of *The Critique of Pure Reason*, the 'A' edition of which was published more than a decade before Kant began writing the cited essay. Alternatively, one might be tempted to discredit Malpas's citation on a cruder basis: as an anachronistic attempt to make Kant look more Heideggerian.

There is, however, something important about Malpas's approach that both these points obscure: his identification of a feature of transcendental inquiry that the approaches of both Kant and Heidegger share as a condition of possibility. This is nothing less than the feature Malpas identifies as 'what is essential to [transcendental] argument as such' (1997: 4); namely, a certain 'circularity of structure'.

Here is how Malpas foregrounds this at the beginning of his essay:

> The main focus of discussion will be the apparent circularity, not merely of transcendental argument, but of transcendental inquiry as such. Such circularity will be taken as presenting, not so much as a problem for the transcendental, as an indication of its essential structure, and of the nature of what Kant called 'transcendental philosophy'. (1997: 2)

As we have seen, Malpas thinks that transcendental approaches need not be constrained to the epistemological paradigm, and that refuting the sceptic is a possible but contingent feature of such approaches. What is necessary for transcendental approaches is the circularity of structure he identifies here. What, then, is the minimum circularity of structure an approach must have in order to be 'transcendental'? On this, Malpas writes:

> The transcendental-ontological project is essentially concerned with 'laying out' a structure that is already present in our being the kinds of beings we are; that is already in the possibility of experience. It does not, and cannot, 'prove' such a structure in any unconditional sense, because the articulation of that structure must itself make essential reference to being as already given, to experience as already presented. ... The project of understanding, conceived from [this] hermeneutic standpoint, is not itself primarily concerned with derivation or proof (though this may well form part of the overall project), but rather with the articulation of a unitary structure within which particular elements can be located and so related to one another and to the whole. (1997: 12)

It is at this point that Malpas's approach is most in danger of seeming to settle for the Heideggerian 'ontological' paradigm for understanding the transcendental, instead of the Kantian epistemological one, as if the choice were a straightforward either/or. This is because the register he adopts here is Heideggerian in its use of terms like 'beings', the 'given' and the 'hermeneutic', and the ontological holism he adopts at the end of the passage is classically Heideggerian, in line with *Being and Time*'s remarks on the 'worldhood of the world' (Heidegger 2005: 91–148).

But what happens if we aim to see past the contingencies of this register, towards the identification of something that exceeds it, and that both the Kantian and the Heideggerian paradigms must share as a condition of their possibility? What, in other words, is the minimum formal condition that allows us to describe a philosophical approach as 'transcendental'?

The answer would seem simply to be this: given X, an approach is 'transcendental' where it enquires into *a priori* conditions for X. This articulation is a more parsimonious version of the definitions of transcendental arguments we considered in Part 2 of this chapter, from Franks and Taylor. This does not mean that it is empty or trivial, however. On the contrary, this apparent formality and emptiness may be precisely what marks out this articulation as the nontrivial condition for describing a philosophical approach as 'transcendental', irrespective of whether that approach subsequently takes on a Kantian 'epistemological' character, a Heideggerian 'ontological' character, or a character that turns out to be irreducible to the presuppositions of either of these approaches.

To draw this out, let me outline five key points associated with the Kantian and Heideggerian transcendental approaches that the minimal definition just offered does not prejudge:

1 Although it presupposes the givenness of 'X', it does not prejudge the mode of this givenness as an 'appearance' in Kant's sense, or an 'entity' in Heidegger's sense.
2 Although it presupposes the possibility of inquiry, it does not prejudge whether this takes place through the agency of a 'subject' as defined by reflexive consciousness in the sense of the post-Cartesian epistemological tradition, or '*Dasein*' in the sense of Heidegger's *Being and Time*-era ontological approach.
3 Although it presupposes that there is circularity involved in the inquiry into the relations between 'X' and its conditions, it does not prejudge whether this is a vicious circularity that is fatal for transcendental arguments, as certain epistemological approaches presuppose it to be (see, for instance, Körner 1966), or whether it is a form of virtuous hermeneutic circularity, as the ontological approaches of Heidegger and Malpas presuppose it to be. All it presupposes, with Malpas, is that it is 'essential circularity of structure' that defines transcendental approaches.
4 Although it presupposes a concept of the *a priori*, it does not prejudge whether this obtains in Kant's sense of the *a priori* as total and timeless, or in the more local and revisable sense of the *a priori* as having to do with the norms structuring 'regional ontologies' that certain commentators have identified in approaches including that of Heidegger (see in particular Friedman 2002; Lafont 2007).
5 Although it presupposes 'conditions', it does not prejudge whether these are 'conditions of possible experience', or what Deleuze, for instance, by way of a critique of the concept of the possible, refers to as 'conditions of real experience' (see Deleuze 1988; Chase and Reynolds 2011: 107). Nor, indeed, does it prejudge whether such conditions must relate to 'possible' or 'real' *experience* at all.[13]

My contention here is that contrasting and emptying Kantian and Heideggerian presuppositions on the transcendental does not leave us with something trivial.[14] Rather, it leaves us with a minimal, formal and nontrivial insight into what might be called transcendental philosophy's properly 'adjectival' form. By this, I mean that it absolves us of the illusion that 'the transcendental' is a noun with either Kantian or Heideggerian properties.[15] Rather, 'transcendental', on the account offered here, is an adjective capable of describing a multitude of philosophical approaches both historical and emergent, with different metaphysical and epistemological commitments.

What 'transcendental' describes, in this sense, are approaches that presuppose some form of *a priori* relation(s) between conditions and what they condition ('X'). What conditions the differences in content between such approaches, in turn, and what marks out the radicalness of their sense of the 'transcendental', is how far they go in problematizing and redefining what is meant by each of the key terms involved in this formulation: What does it mean to 'presuppose' something? What does *a priori* mean? What is a 'relation'? What is a 'condition'? What is 'conditioned' ('X' or 'objecthood')?[16] What, if anything, is 'given', and to what extent is such a concept shrouded in myth/susceptible to critique?

This sense of the transcendental is, I suggest, what is at stake in Malpas's 'circularity of structure'. What this parsimonious sense makes possible is an approach to philosophical inquiry that can be maximally attentive to critiquing the notion of the given ('X'), and dynamic and expansive in tracking its conditions. Viewed in this way, the minimal sense of the transcendental just outlined is not merely circular, but 'recursive'. This is because it involves a capacity to reflect dynamically on conditions that are correlated to an emergent sense of the empirical 'given', and because this process of dynamic reflection has the capacity to generate and make sense of great complexity.[17]

4 Philosophy of technology: Making sense of many turns

Suppose 'X' to be the situation of philosophy of technology today, as evidenced by the tendencies of literature in this field. The aim of this part is to show how the minimal sense of the transcendental just outlined can have implications for a dynamic sense of this field's scope. In doing so, I will seek to defuse objections that could easily be levelled against this minimal sense of the transcendental: that it is, for instance, too theoretical, empty, broad or simply trivial.

I will proceed by developing the three main criticisms of post-empirical turn philosophy of technology outlined at the beginning of this chapter. To recap: 1) that approaches influenced by the empirical turn have tended to repeat a fallacy of reification that they diagnose in 'classical' approaches; 2) that the empirical turn tends towards problematic common-sense presuppositions on what constitutes a 'Technology', to the detriment of the potential for a focus on 'exceptional technologies'; and 3) that the empirical turn has set a problematic precedent where a key picture of method in philosophy of technology is one of 'turning'.

In contrast to the adjectival understanding of the transcendental I have argued for, a dominant tendency in philosophy of technology today is to treat the term as a noun. The reasons for this have to do with the empirical turn that philosophy of technology underwent in the late 1990s.[18] As exemplified in the work of philosophers such as Hans Achterhuis, Philip Brey, Peter-Paul Verbeek and Don Ihde, the empirical turn sought to break with what it saw as the abstract and essentializing tendencies of 'classical' philosophy of technology. Typically, the adjective 'classical' was used in this context to refer to mid-twentieth-century continental approaches taking technology as a theme, and drawing on some mix of psychoanalysis, Marxism, phenomenology, critical theory and hermeneutics. Key figures exemplifying this approach, on the empirical turn account, included Jacques Ellul, Hannah Arendt, Lewis Mumford, Herbert Marcuse, Karl Jaspers, Günther Anders, Erich Fromm, Hans Jonas and, above all, Heidegger.[19]

While the explicit tendency of philosophy of technology since the empirical turn has been to offer mitigated praise of classical approaches (Achterhuis 2001: 3; Verbeek 2005: 7; Brey 2016: 129), a strong implicit tendency for use of the adjective has been pejorative: to homogenize a set of approaches as anachronistic and out of touch with the requirements of the empirical turn. Typically, a number of vices are then levelled against 'classical' approaches, including tendencies towards pessimism, nostalgia, technological determinism and the thesis of 'autonomous technology'.[20]

What approaches in the wake of the empirical turn typically take to be most problematic about 'classical' approaches is a putative tendency to abstract away from case studies of particular artefacts in contexts of design and use, in favour of an essentialism that reifies 'Technology' as a monolithic force, beyond human reason and control (Achterhuis 2001: 1–9; Verbeek 2005, 2011; Kaplan 2009a: 230). Here, for instance, is how Achterhuis characterizes the classical approach at the beginning of the influential 2001 volume, *American Philosophy of Technology: The Empirical Turn*:

> Classical philosophers of technology occupied themselves more with the historical and transcendental conditions that made modern technology possible than with the real changes accompanying the development of a technological culture. Their approach produced invaluable insights. Still, these insights were necessarily circumscribed, because the approach to the technological relation to reality when one is looking at its conditions of possibility leaves unopened many areas of inquiry that can be pursued when one begins to look at the manifold ways in which technology manifests itself. (2001: 3)

On this account, the key mistake of 'classical' approaches amounts to commitment to a sense of the transcendental. The implication is that, where transcendental approaches are indulged, the tendency will be to mistake some set of conditions as 'Technology' itself, at the expense of close attention to how technologies empirically 'manifest themselves' in a variety of different ways and contexts.[21] As Verbeek explicitly puts this point:

> [The] empirical turn constituted a radical shift in approaching technology. It broke away from the predominant focus on the conditions of technology that characterised the early positions. This classical way of thinking can be called 'transcendentalism', because of its kinship to the transcendental-philosophical focus on understanding phenomena in terms of their conditions of possibility. Rather than concentrating on the technological artifacts themselves and their social and cultural impacts, classical positions tended to reduce those artifacts to their conditions, such as the way of disclosing reality they require (Heidegger) or the system of mass production from which they come, which suffocates authentic existence (Jaspers). (2011: 161)

Given the minimal sense of the transcendental I argued for above, a significant problem emerges with Achterhuis and Verbeek's reasoning here. Quite simply, it repeats the fallacy of reification it diagnoses: whereas Achterhuis and Verbeek take classical approaches to reify 'Technology' through transcendental reasoning, they offer a reified vision of the 'Transcendental' as something abstract and out of touch with the empirical in return. To see how this occurs in more precise detail, consider the following passage from Verbeek:

> Traditional philosophy of technology approached its subject matter from a *transcendental* direction. Transcendental philosophy, which achieved its zenith in the work of Immanuel Kant, takes as its point of departure the analysis of conditions of possibility. ... This approach has produced many relevant insights and to a large extent has shaped the understanding of technology and its role in contemporary culture. But our picture of technology is distorted if technology is approached *exclusively* in terms of its conditions of possibility. For then we are speaking about technology's conditions of possibility as if we were speaking about concrete technologies themselves, and the transcendental perspective becomes absolutized into *transcendentalism*. This is precisely what happens in classical philosophy of technology. ... The philosophy of technology needs to resist this 'Orphic temptation' of looking backward. It must be confident that it will be able to get a full view of technology once it has left the realm of the transcendental and re-enters the world of concrete materiality. (2005: 7, Original emphasis)

This passage begins with a strong and fair survey of opportunities and dangers presented by a sense of the transcendental. But note a shift that occurs. At the beginning, 'transcendental' is employed as an adjective to describe 'traditional' approaches to philosophy of technology; by the end of the passage, however, it is employed as a 'realm' abstracted from 'the world of concrete materiality'. This shift is a repetition of the fallacy of reification, and it is problematic for two main reasons. First, it acts against the possibility of reading the transcendental 'adjectivally' and minimally, which is to say, not as a static realm of essences out of touch with the empirical, but rather as a circular and recursive approach to argument or method that can be dynamically in touch with the minutiae of the empirical. Second, it acts against the possibility of more fine-grained interaction with philosophers who employ this sense of the transcendental, including thinkers from the broader 'continental' philosophical tradition.[22]

To get a clearer sense of why these issues matter, note Verbeek's concluding emphasis on 'the world of concrete materiality', and Achterhuis's emphasis on a set of 'real changes accompanying the development of a technological culture'. Adjectives such as 'concrete' and 'real' are used here to endorse the credentials of the empirical turn, and to discredit previous 'classical'/ 'transcendental' approaches. So far in this part, I have argued that this reasoning involves a fallacy of reification. Beyond this, however, a further problem emerges when we consider the kind of limits that the notion of an empirical turn sets on what constitutes an object worthy of inquiry.

Robert C. Scharff has formulated a version of the point touched upon here as follows:

> Advocates of [the empirical] turn are never very clear about what 'empirical' rules in … but they are very clear about what it leaves out. Any talk about Technology *Überhaupt* will either take us back to the bad old days of metaphysical pronouncements about unchanging essences, moral and political absolutes, and experience ignoring accounts of what is 'really' going on, or force us to listen to social scientific explanations about how things got this way and, barring new 'conditions', must therefore continue this way. (2012: 154)

The problem that Scharff identifies here on what constitutes the 'empirical' is far from trivial. This is because, where it goes unaddressed, it leaves philosophy of technology dependent on external and unclarified standards of 'self-evidence' and common sense as its most obvious authorities on this matter. Such standards may themselves be influenced by one of various forms of empiricism in the epistemological tradition (whether Humean, Comtean or Carnapian, for instance), or, more problematically, by the most recent artefacts produced by industry. Either way, a key danger emerges for the scope of philosophy of technology as a field: a tendency to drift towards narrow forms of positivism and presentism that prioritize case studies of what might be called 'zeitgeist-seizing' technologies.[23] Such case studies may be selected because of their common-sense utility, their visibility, their perceived timeliness or their ability to attract and sustain funding. The point, however, is that where attention is directed towards them, we stand to become inattentive to more marginal and 'exceptional' technologies that trouble our common-sense conceptions of the empirical, but that might be just as (or *more*) significant for addressing problems and issues concerning technology across a range of different contemporary contexts (whether, for instance, political, epistemological, economic, or engineering and design-focused).

One way to draw out the stakes of this issue is to historicize our consideration of what constitutes the empirical. As David M. Kaplan has put it in a review of Verbeek's approach:

> Verbeek makes it seem as if all backward-looking approaches to technology are transcendental. They are not. There are backward-looking approaches that are instead historical and material. The best known example is 'historical-materialism', a theory that maintains that persons, events, and things are best understood in relation to their historical development. (2009a: 234)[24]

The approach Kaplan advocates here brings the advantage of connecting to a wide body of Marxist and hermeneutic thought on what constitutes 'the empirical' (see, for instance, Jameson 2002; Ricoeur 1984; Gadamer 2004; Ihde 2007: 112–13). However, can't we open still wider resources by changing the emphasis of his point? On the formal view of the transcendental outlined in this chapter, 'backward-looking' historical and material approaches to conditions are in fact forms of 'transcendental' inquiry par excellence.[25]

The point I want to emphasize here is this: Historical approaches can broaden our field of attention from a narrow focus on zeitgeist-seizing artefacts, in favour of case studies of historical ones; similarly, materialist approaches can broaden our sense of things to include a focus on the material culture or socio-economic conditions shaping the emergence of particular artefacts and practices; when we contextualize historical and materialist approaches as part of a broader tradition involving a 'transcendental' approach to argument or method, however, the scope of inquiry can be broadened even further, to include attention to what I have called 'exceptional technologies'.

As outlined in the introduction to this book, 'exceptional technologies' are artefacts and practices that appear as marginal or paradoxical exceptions to a received sense of what empirically constitutes a technology in a given context of design, implementation or use, but that can nevertheless act as important focal points for drawing out and challenging conditions implicated in the received sense.[26] My claim is that ostensibly trivial, failed, merely imagined and impossible technologies can all be 'exceptional' in this sense. This is because these artefacts and practices can, for instance, be implicated in setting the conditions for funding, expectation, and the hopes and fears that are invested into technologies, as well as the differences and similarities between how artefacts and practices figure across diverse social imaginaries and moral codes. As a corollary to this, my claim, to be developed over the course of this book as a whole, is that the minimal sense of the transcendental outlined in this chapter can act as a way of being open to focusing on such exceptions. This is because it allows us to draw on and compare a range of different philosophical approaches to the conditions that constitute our received sense of the empirical, and, in doing so, to track what this sense renders 'normal' and 'exceptional' in a dynamic and evolving way.

As a concluding point for this chapter, let me emphasize that what I have just argued for does not amount to a call for a new 'turn' in philosophy of technology. Since the empirical turn, there have been many calls for further turns that act on its precedent, including, for instance, an 'engineering turn' (Vermaas 2016), 'ethical' or 'axiological' turns (Verbeek 2011; Kroes and Meijers 2016), a 'societal turn' (Brey 2016), a 'semantic turn' (Krippendorff 2005), a 'practice turn' (Hillerbrand and Roeser 2016), a 'narrative turn' (Kaplan 2009b) and a 'policy turn' (Briggle 2016).

Rather than advocating a further turn, the point is this: what may be required in philosophy of technology today is a more thoroughgoing consideration of the conditions under which such 'turns' are possible.

This move would not amount to a 'transcendental turn' in all but name, because it would be aware of, and resistant to, the tendency to reify 'Technology' that thinkers like Achterhuis and Verbeek are right to be cautious of in 'classical' approaches to philosophy of technology. And nor would it be some form of 'ontological turn' running parallel to recent developments in anthropology (see Descola 2013). Instead, it would be a movement further in both 'transcendental' and 'empirical' directions at once: towards as broad, de-reified and minimal a sense of the transcendental as possible, and towards as maximally attuned and fine-grained a sense of the empirical as possible, according to the requirements of a given problem ('X').

It might immediately be objected that it is ridiculous to try to move in more than one direction at once. However, as I will develop over the course of this book (and in Chapter 5 in particular), it might be that this only seems to be the case when we ascribe to a crudely 'first-person' picture of 'turning', where every 'turn towards' seems to necessitate a 'turn away from'.

The precedent instance of this in philosophy of technology, as I have claimed in this part, has been the empirical turn away from 'classical'/'transcendental' approaches. What I will suggest as an alternative in this book can (but need not) be framed in terms of different pictures: of 'topography', 'topology' or 'mapping'.[27] In each of these cases, what is aimed at is a dynamic but non-reductive sense of the field on which philosophy of technology's various 'turns' are taking place. This involves a shift 'underneath' the picture of turning, as it were: not to disqualify it, but to contextualize it in terms of an evolving sense of its conditions, and to mitigate the sense of disorientation and fragmentation that the proliferation of turns taking place since the empirical turn may have led to. To take spatial metaphors further, what is being sought here is analogous to the shift from a first-person sense of direction in space to a survey of the space itself. This is important, because, whereas every 'turn towards' involves a 'turning away from' for the former sense, it is perfectly possible to move in more than one direction for the latter, at varying levels of complexity: by expanding and contracting the scope of the space surveyed by zooming in and out.

CHAPTER TWO

The blank page

In Chapter 1, I argued that a renewed sense of the transcendental might make sense for philosophy of technology today. This would involve going further, at once, in directions that the legacy of the empirical turn in this field has tended to position as mutually exclusive. In a 'transcendental' direction, the aim would be to de-mystify and de-reify our sense of the transcendental so as to open philosophy of technology to different perspectives on methods, thinkers and concepts drawn especially (but by no means exclusively) from the continental tradition in philosophy.[1] In an empirical direction, the aim would be to mitigate the potential for philosophical considerations of technology to drift towards an exclusive and positivistic focus, in favour of a developed concept of 'exceptional' technologies.

A big objection that might immediately be levelled against this approach concerns relevance. Even if we grant some merit to expanding philosophy of technology's sense of the transcendental, how can this help us decide when any inquiry into a given artefact, practice or problem, 'zeitgeist-seizing', 'exceptional' or otherwise, has been conducted to an appropriate degree? What, in other words, is there to stop the sense of the transcendental advocated in this book from sliding into an infinite regress of conditions?

The aim of this chapter is to respond to this problem through a case study of an exceptional technology. The study in question concerns the 'blank page', understood in a sense broad enough to cover, for instance, the literal sense of a sheet of paper, the figurative sense of Locke's famous tabula rasa, and senses current in the use of new media that take up and figuratively 'remediate' this notion, such as that of the blank address bar of an internet browser, or the blank field of a search engine.[2]

Choosing to focus on this case study may seem simply to compound the problem of relevance. It may, for instance, seem too broad to be properly 'empirical', and it may seem to speak to the highly constrained practical outlook of philosophers, for whom, with the notorious exception of Socrates (see Derrida 2016: 6), 'blank pages' have invariably featured as technological conditions for the working out, writing and recording of arguments and concepts. At the limit, this might make this chapter seem to tend towards the worst aspects of what has been called 'classical' or 'humanities' philosophy of technology, armchair speculation or, indeed, a form of 'wine tasting' aestheticism (see Briggle 2016: 168–9).[3]

It is, in fact, for precisely these reasons that the blank page has been chosen as this chapter's focus. This is because it problematizes our concept of what can count as a relevant and empirically oriented case study in philosophy of technology, and because working through the complexities of its shifting senses will, as I hope to show, allow us to develop, apply and enact the renewed sense of the transcendental argued for in Chapter 1. Stated differently, the contention of this chapter is this: the 'blank page' is an example of an exceptional technology that hides in plain sight across diverse cultures and practices, and that requires a form of transcendentally informed inquiry to think through its complexities.

The aim of this chapter, then, is to address the problem of relevance by showing how the potential for an infinite regress can be contained for the sense of the transcendental developed in this book. For Kant and many thinkers in the transcendental tradition, what ultimately contains such a regress is a bedrock form of subjectivity.[4] What contains it on the account developed in this chapter is attention to the exceptional demands of a given technological artefact, practice or problem.

In Part 1, I focus on how the blank page operates as a theme across two of the phenomenologist Edmund Husserl's most influential works: *Ideas* and *The Crisis of European Sciences and Transcendental Phenomenology*.[5] The aim here is to draw out the sense in which Husserl's late 'transcendental phenomenology' implicates more complex material and historical conditions than its focus on subjectivity as the 'primal source' of meaning might lead us to suppose (Husserl 1970: 99). It is acknowledged that such a reading is far from novel in the continental tradition; what is novel, however, is a focus on the theme of the blank page as a way of enacting it. This is because it allows us to practice a form of what Husserl called 'imaginative variation' on Husserl's own practice of writing.[6] What emerges from this, I hope, is a reading of the blank page that runs counter to any received sense of it as a trivial object, unworthy of sustained philosophical inquiry. Instead, I argue that the blank page is an exceptional technology, implicating diverse and complex conditions. This part concludes by contrasting my approach with a reading of Husserl's 'missing technologies' from Don Ihde (2016).

Part 2 considers how conditions implicated by the blank page have changed historically since Husserl's time. Today, for instance, we can imagine applying the term 'blank page' to all manner of software and hardware innovations of computing, from word processor documents and the fields of a search engine, to tablet and smartphone touch screens. I argue that this extension, while apparently trivial, in fact offers an important example of how figurative treatment of an artefact can obscure and simplify complexities in the conditions that constitute different artefacts. I seek to demonstrate this by extending the practice of imaginative variation argued for in Part 1 to contrast Husserl's situation with two contemporary situations involving different 'blank pages': taking a written exam and using a search engine.

Parts 3 and 4 argue that the extended form of imaginative variation practised in this chapter can act as a 'continental' complement to recent attempts to extend the scope and method of philosophy of technology that draw on the analytic tradition. Part 3 enacts three further short variations that draw on the respective theoretical frameworks of Husserl, Heidegger and Deleuze to show how imaginative variation can be framed as an enacted and situated way of meeting the problem of relevance facing a sense of the transcendental. Part 4 then concludes with a critical reading of the approaches of Franssen and Koller (2016) and Franssen et al. (2013). I argue that while these approaches rightly emphasize the capacity of the analytic tradition to extend the scope and method of philosophy of technology today, they arbitrarily exclude approaches drawing on 'social science and humanities' (Franssen et al. 2013), and overlook the merits of a sense of the transcendental emerging from the continental tradition.

1 'This white paper'

In Section 35 of *Ideas*, Husserl writes:

> Let us start with an example. In front of me, in the dim light, lies this white paper. I see it, I touch it. This perceptual seeing and touching of the paper ..., precisely with this relative lack of clearness, with this imperfect definition, appearing to me from this particular angle – is a *cogitatio*, a conscious experience. (2002: 65)

Husserl seems to be doing something extremely straightforward and classically phenomenological here: closely describing what it is like to be conscious of something that is 'given' to consciousness. This thing, it seems, need not have been a piece of paper; it could, with Heidegger, have been a hammer, or, with Sartre, a glass of beer or a Chestnut tree (Heidegger 2005: 107–14; Sartre 2008: 20–6, 180–92). But what if we pose a speculative question that adds another layer of complexity, and that touches on conditions relevant to Husserl's practice: in what sense is Husserl conscious of the paper before him as an example?[7]

How conscious, in other words, was Husserl in this instance that a piece of paper has particular qualities that render it immediate, accessible and dramatic as an example for the unfolding of a philosophical position? Husserl may, for instance, have consciously selected the paper because he presupposed that the reader, in the act of following the example, would also be in intimate contact with a piece of paper.[8] But perhaps we've got it wrong. Perhaps Husserl was more or less unconscious of the qualities of the paper before him. Perhaps he simply paused during the act of writing, looked up and unimaginatively seized on a piece of paper as, in Heidegger's terms, one of the most 'ready-to-hand' (*Zuhanden*) aspects of his immediate surrounding environment or '*Umwelt*' as a writer.

We can go much further in this speculative direction. Suppose that Husserl is sitting before the first draft of section 35 of *Ideas*. He has just finished the sentence immediately before the one introducing the paper example, but is now encountering a bad case of 'writer's block'. Husserl knows he wants to use an example, but he wants a good one that will engage the reader, and is struggling to come up with it. He is conscious that what he has written so far is no guarantee of his ability to proceed. He is conscious of the space remaining blank on the page. He is becoming ever more conscious that *Ideas* is, at this moment, only potentially a publishable piece of writing, and he is becoming anxious that it might never actually become one.

Many more such 'imaginative variations' are possible. We might, for instance, imagine that Husserl has just arrived at the idea of using the paper example, and that his situation has switched from one of writer's block to a situation of what we might call 'writer's excess': Husserl now has lots of good ideas, but does not know how to order them on the page. We might alternatively imagine that he is several drafts away from the example, and has just torn up an early draft of section 35 in frustration. We might imagine that he has completed the entire first draft of *Ideas*, and is now using it to prepare a final manuscript for the publisher, on a typewriter.[9] We might imagine that, like Hegel at Jena in 1806, he is about to dispatch his only copy of the manuscript, under fraught conditions. We might imagine an indefinite array of historically false or improbable situations: that Frege wrote *Ideas*, or that Husserl wrote it in the 1970s using a word processor.

What worthwhile philosophical purpose can such variations serve? At a minimum, the claim I want to develop in this part is that they provide a way of foregrounding and exploring what it was for Husserl to be caught up in writing as a process with complex technological conditions. Viewed in this way, the question posed above concerning Husserl's consciousness of the paper turns out to have been highly rhetorical (and not at all a fetishistic instance of the 'intentional fallacy'). This is because we can readily imagine that, as someone implicated in the process of writing, it was necessary for Husserl to be conscious of the paper as a technological artefact, not merely in terms of the '*cogitatio*' he describes, but in terms of multiple and shifting '*cogitatios*'. When immersed in the creative process, Husserl approaches the paper as a field full of possibilities. When doubtful, he does not know how to fill it. When frustrated, he wants to destroy it. When revising or dispatching it, he takes every care that it is preserved. The point is that in each of these cases, and in indefinitely many others, we can readily imagine Husserl being acutely conscious of the paper before him, and in each case we come back to a key material condition for the possibility of this: that Husserl must presuppose this paper to be an artefact adequate to facilitating the working out, recording and dissemination of his ideas.

That we can imagine Husserl engaged in writing in this way provides an important alternative perspective on his own attempts to arrive at what he called 'transcendental phenomenology'. By this, Husserl meant a form of methodologically first philosophy that took up and radicalized Kant's approach to the transcendental by supplementing it with a different conception of the constitutive role of consciousness (see Husserl 1988; Zahavi 2003, 2008). As Husserl writes in *The Crisis*:

> Should I, in the following presentations, succeed – as I hope – in awakening the insight that a transcendental philosophy is the more genuine, and better fulfils its vocation as philosophy, the more radical it is and, finally, that it comes to its actual and true existence, to its actual and true beginning, only when the philosopher has penetrated to a clear understanding of himself as the subjectivity functioning as primal source, we should still have to recognise ... that Kant's philosophy is on the *way* to this. (1970: 99, Original emphasis)

Here, as throughout *The Crisis*, Husserl presupposes subjectivity to be the methodologically key condition for radicalizing transcendental philosophy beyond Kant (see also Husserl 1970: 68–9). In *Ideas*, this is, for instance, played out by offering new concepts of intuition and the dynamics of time consciousness (2002: 39–40, 230–3). In *The Crisis*, it is played out most famously through Husserl's development of a concept of the 'Lifeworld' or *Lebenswelt* as the 'pregiven' condition for all subjective endeavours in science (1970: 48–53).

By virtue of the subjective emphasis of these presuppositions, a common critique has developed against Husserl in the subsequent continental tradition: that his approach remains indebted to a sense of the subject that is abstract, and that it overlooks more fundamental conditions for its own possibility (Ihde 2016: 47; 2012: 117; see also Dreyfus 1991). This line of critique most famously begins with Heidegger's positioning of Husserlian phenomenology as a 'method' on the way to 'fundamental ontology' (2005: 62–3, 73–4). It is also a feature of many other key readings, however: from Sartre's critique of the Husserlian concept of the ego as psychologically individuated in *Transcendence of the Ego* (1972), to Deleuze's development of this Sartrean position in *Logic of Sense* (2004b: 112–13), to Merleau-Ponty's development of comments in Husserl's late work on embodiment (1976), through to Derrida's reading of 'Difference' as an obscured condition for Husserl's work (1973, 1989: 153).

In terms of contemporary philosophy of technology, it is also possible to situate Don Ihde's career-long engagement with Husserl as part of this line of critique.[10] Consider, for instance, the following from Ihde on a famous characterization of Galileo that occurs in Husserl's *Crisis*-era work:

> The science/lifeworld distance that Husserl claims originates in Galileo is admittedly enhanced by some of Galileo's own rhetoric but not by his praxis. ... Husserl is also 'forgetful' insofar as he ignores the transformational mediation of the telescope within Galileo's praxis. (2016: 55–6)

In *The Crisis* and Husserl's related 'Origin of Geometry' essay, the proper name 'Galileo' comes to symbolize the beginnings of a split between the Lifeworld as presupposed by everyday experience and the world of modern science as a derivative and 'mathematized' abstraction (Husserl 1970: 23–4). Ihde's point in the above passage is that no such split is possible in principle, and that all science, no matter how abstract, must retain traces of its Lifeworld conditions (Ihde 2012: 115–28).

What must be noted here is that this point is also one of the key motivations behind Husserl's *Crisis*. Ihde's claim, however, is that Husserl cannot adequately elaborate the point because of limits inherent to his practice. As a philosopher trained to focus on the pronouncements that Galileo makes in his theoretical writings, Husserl, is, Ihde thinks, destined to overlook the forms of praxis that condition Galileo's mode of being-in-the-world as a scientist. These forms are crystallized for Ihde in this case by Galileo's telescope, considered as an artefact that enables extended modes of visual perception (Ihde 2016: 13). Correlatively, Ihde thinks that it is Galileo's practical interactions with such artefacts that should be the primary object of Husserl's inquiry (2016: 55).

For these reasons, Ihde finds it difficult to situate Husserl as a 'philosopher of technology'. Commenting on Husserl's examples, he notes:

> What types of examples do we find in Husserl? First, psychological, particularly perceptual-psychological examples abound to good use: Listening to musical tones, tactile examples from the hand, memory examples of pretension and retension, certain visual examples – all abound to good if usually brief purpose. Then, there are the objects-before-one examples, and ... carpentry examples of smoothing and shaping. But – and this is a heuristic question – where are the instruments? The tools? The artifacts that are productive of change, insight, or transformation? (2016: 52)

To highlight what is at stake here for Ihde, consider the differences between Husserl's protagonist 'Galileo' and Ihde's protagonist 'Husserl': whereas the example of the telescope strikes Ihde as an obvious feature of Galileo's praxis that Husserl overlooks, no comparable praxis-oriented example strikes Ihde as readily available to explore the technological conditions involved in Husserl's praxis. At the limit, however, he notes one apparent exception to this (although it too will ultimately turn out to frustrate him):

> [Writing] is perhaps as close as Husserl comes to identifying a material technology and its praxis playing a role in which meaning-structures are not alienated from lifeworld praxis. ... It would be stretching Husserl to claim that this is much of a recognition of technological artifactuality in a mediating role in any very detailed way, but at least there is a recognition that materiality can, through its very material transformation, make meaning-structures available to bodily humans. (2016: 53)

At the beginning of this part, I stated a minimum aim: to use imaginative variations on the theme of the blank page as a way of foregrounding and exploring a sense of what it was for Husserl to be caught up in writing as a process with complex technological conditions. To conclude this part, let me attempt to highlight some of these conditions by contrasting this approach with Ihde's reading of Husserl.

Ihde asks: 'Where are [Husserl's] instruments? The tools? The artifacts that are productive of change, insight or transformation?' The type of imaginative variation practised in this part suggests that Ihde's questions overlook something important here: that blank pages must consistently feature as precisely such 'instruments', 'tools' or 'artifacts', not merely for Husserl's practice of writing, but also for Ihde's practice of reading Husserl. Without his 'white paper', Husserl would have lacked a necessary condition for the working out, recording and transmission of his philosophy, including all the examples Ihde cites above. Reciprocally, without the blank spaces surrounding and transmitting Husserl's words to him, Ihde would lack a necessary condition for his reception and interpretation of Husserl's philosophy, and this irrespective of whether he might be reading it in manuscript form in Husserl's archive, in a published German or English paper edition, or in a digital format.[11]

When viewed in this way, 'blank pages' emerge as a necessary technological condition for producing precisely the types of change, insight and transformation Ihde's question seeks, both for Husserl as a writer and for Ihde as a reader, and in multiple ways.

Materially, blank pages must be in place as the substrate allowing any letter, then word, then sentence to become articulated with any other such unit across Husserl's writings. But it is also the case in diverse other ways too. Phenomenologically, the page that appears as 'blank' must play at least as important a role as the marks made upon it for making the experience of reading Husserl possible. Historically, the marks made upon the pages of Husserl's texts may decay, corrupt or become unfashionable, thereby fading his writings back to 'blankness' in multiple literal and figurative senses. Ecologically, paper versions of Husserl's texts may be pulped then recycled: radical transformations that will cause any putative insights they contain to be changed entirely, but which will make their 'blankness' re-emerge in different ways, for potential re-incorporation into other artefacts.[12]

Ihde's reading of Husserl overlooks these complex implications of the blank page because it seeks an explicitly formulated set of examples of writing technologies. This is what leads Ihde to comment that 'it would be stretching Husserl to claim that [his work on writing is] much of a recognition of technological artifactuality in a mediating role in any very detailed way'. In contrast, the type of imaginative variation practised in this part suggests that Ihde's approach is not 'stretched' enough. This is because it overlooks the sense in which Husserl's entire written *oeuvre* is exemplary, in a series of very detailed ways, of the complexities of 'technological artifactuality in a mediating role'.

As Ihde notes, Husserl's *oeuvre* contains a relative dearth of explicitly cited examples focused on what we might intuitively take to be the empirically verifiable writing technologies of his day, such as pens, typewriters and desks (Ihde 2016: 59–76). Ihde does not comment, however, on the status of the blank page in Husserl's work, considered as a more counterintuitive and 'exceptional' writing technology. Had Ihde expanded his inquiry to include this artefact, however, he might similarly have been frustrated. This is because blank pages likewise figure relatively infrequently across Husserl's writings as an explicitly cited example, despite the example of 'white paper' cited at the beginning of this part.

This noted, the type of imaginative variation practised in this part has aimed at two broader points: first, the writing technologies that Ihde seeks are all implicated in Husserl's work, not as theoretical objects in the form of examples (the instances of which might be empirically catalogued), but as necessary practical conditions for the possibility of the work; second, a focus on blank pages, considered as a more counterintuitive and 'exceptional' writing technology, offers a way of enlivening our sense for the complexities of this implication.

2 Varying conditions

In focusing on Husserl's 'white paper', have we spent too much time discussing something that is simply trivial, rather than a so-called exceptional technology? Alternatively, have we placed too much emphasis on an example that seems to imply some form of metaphysically pure 'blankness'? To counter both these worries, this part will extend the practice of imaginative variation I have argued for beyond Husserl's situation, to contrast it with two contemporary 'blank page' situations. The aim in doing so is to address two questions: what is it to be faced with a 'blank page' today, and in what ways have conditions relevant to such situations changed from those relevant to Husserl?

Picture a contemporary situation where being faced with a blank page is an isolating experience. Take the situation of a student who has just opened a script book in a written exam. The book is blank, and they are expected to fill it. Irrespective of whether they have prepared well, and of whether they are confident or nervous, this situation is contrived to isolate the student: it is a disciplinary '*dispositif*' in Foucault's sense, engineered to isolate and test certain key competences in the individual, such as memory, time-keeping and equanimity (Foucault 1991).

How does this isolation differ from that imaginable in the case of Husserl's 'writer's block', as discussed above, or from that imaginable in the case of a student taking a written exam in Husserl's time? To address this, we have to consider a material change that has occurred in the conditions under which the contemporary student's situation can be imagined: while it is by no means universally the case for a variety of often vexed economic, geographical or pedagogical reasons, we can imagine this student conducting most of their writing on devices networked to the internet.[13] Now removed by exam conditions, these will have featured as significant background conditions for the student's normal written output. In contrast, Husserl and his contemporaries did not encounter this same dichotomy between 'exam' and 'normal' conditions of writing to anything like the same degree.[14]

Now imagine a situation where being faced with a blank page seems to be a 'connected' experience. Take the situation of watching the cursor blink on a search engine. The search field is blank, and the user has indefinite time to fill it. Irrespective of whether the user is clear on what they want to search for, this situation has been contrived to connect them: at its heart, a search engine lies ready to facilitate and record the user's abilities to access vast databases of information.

How does this connectedness differ from that imaginable in the case of what we called Husserl's 'writer's excess'? To address this, we have to consider a figurative shift in the conditions for posing the question: we have now extended the sense of the term 'blank page' to include internet search engines. The key point, however, is that such an extension *makes sense* given the technological changes noted above. Today, our concept of the 'blank page' functions much more figuratively than it did in Husserl's time, and this, in large part, is due to innovations in post-World War Two computing, which have extended the concept to include a plurality of different artefacts and platforms.[15]

Today, it makes sense to refer to word processor documents, command and search engine prompts, browser address bars, monitors and touch screens as 'blank pages', across contexts that are more or less 'everyday' and 'specialized'. Technological change therefore seems to be prompting a more abstract definition of the term: a 'blank page', it seems, is any apparently empty or blank medium capable of being filled with information.

This definition appears abstract enough to cover all imaginable literal and figurative senses of a 'blank page'. What it hides, however, are precisely the different conditions that constitute different types of 'blank page'. To make this clear, let us attend to one word in the definition in particular: 'apparently'.

Almost invariably, the interface greeting us when we connect to a contemporary search engine exemplifies clear and smooth design principles. Trivially, this is because companies investing in the engine have an interest in making it as 'user-friendly' and ergonomic as possible. More importantly, however, what the interface enshrines is a kind of 'principle of principles' governing the culture of ubiquitous computing more generally: that apparent simplicity at the level of the interface should mask massive complexity at the level of the coding and hardware infrastructure.[16] An internet search engine is therefore only ever seemingly a 'blank page'. In fact, it may be one of the most highly coded 'spaces' the contemporary internet user is likely to encounter, whether online or offline (see Hayles 2012: 6).

The point here is by no means to set up Husserl's white paper or our imagined student's exam script as pure and 'uncodified' spaces, in contrast to the highly coded space of the search engine. On the contrary, it is to think through contrasts between these three artefacts to show just how highly coded each are by their conditions, in different ways.

Like the search engine, Husserl's white paper and the student's script book lie ready to facilitate and record the abilities of their users according to all manner of complex conditions, including the types of material, phenomenological, historical and ecological conditions I considered towards the end of Part 1 above. Beyond these, however, there are multiple other relevant conditions that encode Husserl and the student's possible modes of engagement with their respective 'blank pages'. These include different grammatical, orthographical, lexical and poetic conditions, as well as codes of perspective and draughtsmanship (should we imagine Husserl, himself under the sway of imagination and reverie, or our imagined student, perhaps from frustration, deciding to draw rather than write on their respective pieces of paper (see Bachelard 1970: 31)). In contrast, a different set of conditions is relevant to the situation of the search engine user, which we might imagine as follows: they want information they take to be more or less publicly available; they have delegated search capacity to the engine; they expect relevant information to be filtered from redundant information; and, in contrast to the situation of Husserl and the student, they expect the whole process to involve minimal subjective effort.[17]

What the expanded type of imaginative variation practised in this part offers, I suggest, is a way of highlighting and problematizing relevant differences in conditions across situations like these, and, thereby, of complementing (but not replacing) thoroughgoing empirical inquiry into these situations. In doing so, it should be noted that it runs counter to two tendencies. First, it runs counter to a tendency in contemporary philosophy of technology to overlook the complexity of conditions implicated by exceptional technologies. Second, it runs counter to what, following Wittgenstein, we might call the 'grammatical' tendency of diverse language games, both everyday and specialized, to treat technologies figuratively, in ways that abstract from and distort the complexity of conditions shaping the artefacts and practices in question (Wittgenstein 2009).

In contrast, the type of imaginative variation outlined in this part highlights two important points. First, no 'blank page' is ever blank. Rather, whatever appears as a blank page, whether in a literal or a figurative sense, is in fact a complex artefact condensing a range of historically shifting conditions that go into constituting its imaginable set of uses. Second, and correlatively, there are many different 'blank pages', each with a complex set of conditions.

Let me now conclude this part by considering two important objections. First, it might be objected that the points I have just made, as per the whole focus of this chapter, belong to a dated 'textual' paradigm in continental philosophy that fetishizes texts and writerly practices (see James 2012: 1–16). Second, it might be objected that they make 'blank pages' so singular as to make it impractical to inquire empirically into the conditions implicated by any such artefact.

In response to the first objection, it should be emphasized that the case of the blank page highlights something about the figurative treatment of technologies in general: that such extension tends to abstract from and simplify a focus on conditions. What imaginative variation in the case of the blank page exemplifies, in contrast, is an enacted and contextually constrained sense of the transcendental that is critically attentive to conditions, and that might be applied in philosophy of technology to analyse the consequences of figurative treatments of technologies per se, by exploring conditions relevant to notorious historical cases such as the metaphors of the body as a 'machine' (see Descartes 2006; La Mettrie 2006; Kang 2011), through to more contemporary metaphors that frame the mind as 'software' and Being as 'information' (see Hayles 1999, 2012; Floridi 2013).

In response to the second objection, it should be emphasized that the type of imaginative variation practised in this part is suggested as a potential complement to detailed empirical work in and across diverse contemporary fields, and not as a replacement. Instead of undermining the empirical practices of ecologists interested in the consequences of a 'paperless economy', of psychologists or neuroscientists interested in exam stress, or of sociologists interested in evolving patterns of search engine use, then, the aim is this: to highlight the sense in which imaginative variation involves an enacted sense of the transcendental that is always situated, and that may be useful for exploring different research contexts.

3 Re-imagining relevance (1)

How do things stand in philosophy of technology today with respect to the point just made? In other words, how might philosophy of technology after the empirical turn take up a renewed sense of imaginative variation as a situated and enacted practice? And how might it do this in a way that complements and extends its work on case studies while also addressing the problem of relevance, as outlined at the beginning of this chapter?

To focus these issues, consider the following summary from Maarten Franssen and Stefan Koller, two thinkers representative of an emergent strain of Dutch philosophy of technology that draws on the analytic philosophical tradition:

> Notwithstanding the empirical turn, work in the philosophy of technology is still too fragmented and isolated, both internally, in how its various themes are mutually related, and externally, in how well its themes are linked up to what happens in the major fields that make up philosophy as a whole. We argue that ... philosophy of technology as currently practiced has to extend both in scope and method and that a systematic exploration of its connections with the core fields of philosophy will help it develop into a mature field. ... Greater systematicity is *needed* to counteract the fragmentation and lack of substantive unity in philosophy of technology. Such systematicity can be *provided* by ... checking the content and validity of new contributions against both extant results in philosophy of technology and (conceptually or inferentially) related positions in foundational analytic philosophy, above all metaphysics, epistemology, and the philosophy of language. (2016: 31, Original emphasis)

The first thing to note here is that Franssen and Koller share an explicitly stated aim with the approach developed in this book: to extend philosophy of technology 'both in scope and method'. To do so, however, they recommend drawing on a different philosophical tradition: foundational analytic philosophy, instead of 'transcendental' continental philosophy.

On an uncharitable and partisan reading of modern philosophy's analytic/continental divide, the second point I just made might be taken to disqualify the first. In other words, if one is drawing on the continental tradition, what one aims at in terms of 'scope and method' will turn out to be something very different from, and incompatible with, what one aims at if one is drawing on foundational analytic philosophy. Examining the themes of this chapter in relation to those of Franssen and Koller, this point would seem to be borne out in practice: whereas this chapter has developed a series of historically scattered imaginative variations focused on the blank page as an 'exceptional technology', Franssen and Koller seem to aim at something very different: an outline of prospects for a 'unified philosophy of technology' focused on taking the 'activities of engineers and the conceptual and tangible outputs of these activities ... seriously' (2016: 32–3).

A more charitable reading than this is possible, however. Indeed, it is something that Franssen and Koller leave scope for when claiming that analytic systematicity 'can' be a way of developing philosophy of technology, not that it *must*. On a more charitable reading, this implies that appeal to foundational analytic philosophy is not the only means for extending philosophy of technology as currently practised. Rather, there may be different means to a similar end, and these can be complementary rather than mutually exclusive. This is simply to recognize something that ought, as I will argue below, to be trivial: different approaches can and should contribute to the development of philosophy of technology today, regardless of whether their background draws on the continental or analytic traditions, or, indeed, from other traditions and fields of thought.

Reconsider some of the post-Husserlian thinkers in the continental tradition I cited in Part 2: Heidegger, Sartre, Deleuze, Merleau-Ponty, Derrida, Ihde. I claimed that each of these thinkers can, in different ways, be viewed as developing a common line of transcendental critique against Husserl: that his approach remains indebted to a sense of the subject that is abstract, and that it overlooks more fundamental conditions of its own possibility. As key Husserl scholars have argued, this line of critique can often seem to be based on a straw man image of Husserl that overlooks significant details of the fundamentally Lifeworld-oriented nature of his later work (Zahavi 2003: 43; Moran 2000, 2012). If this point is taken to be sufficient to block this line of critique, however, something else gets ignored: that the critique can, by virtue of its recurring focus on overlooked conditions, also be viewed in terms of a shared and ongoing project to explore and develop variations on a sense of the transcendental.

My suggestion here is this: by being open to the sense of such a shared project, philosophy of technology after the empirical turn can set in place conditions for drawing more extensively and creatively on the continental tradition than has been the case up to now, and in ways that can (but need not) complement approaches premised on foundational analytic philosophy, such as highlighted by Franssen and Koller.

There is a sense in which this approach might be complementary that should be emphasized immediately: a crucial dimension of Franssen and Koller's focus on metaphysics, epistemology and philosophy of language is that it has the potential to free philosophy of technology from any recalcitrant sense that it must limit itself to drawing on the work of philosophers who have treated technologies as an explicit theme.[18] This sense, arguably more practised in philosophy of technology than explicitly avowed, is in fact detrimental for extending the scope and method of the field. This is because it restricts scope to drawing on a small subset of all available philosophical positions, and because it pushes method towards something akin to a scavenging exercise focused on finding explicitly cited examples of technologies in the texts of philosophers.

What I want to suggest is that openness to a sense of the transcendental that is prevalent in the continental tradition has the potential to open philosophy of technology to a wider range of philosophical methods than this, and, by consequence, to a wider range of technological artefacts, practices and problems.

What is particularly important to note in this respect is that this approach is in fact consistent with Husserl's own later sense of the transcendental. Writing in *The Crisis*, he notes:

> When we proceed, philosophizing with Kant, not by starting from his beginning and moving forward in his paths but by inquiring back into what was thus taken for granted ..., there opens up to us, to our growing astonishment, an infinity of ever new phenomena belonging to a new dimension, coming to light only through consistent [inquiry into] what was ... taken for granted – an infinity, because continued penetration shows that every phenomenon attained through this unfolding of meaning, given at first in the life-world as obviously existing, itself contains ... implications whose exposition leads again to new phenomena, and so on. (1970: 111)

Situated in terms of this passage, a key aim involving a sense of the transcendental emerges: to find new ways of exploring what Husserl refers to as the 'infinity of ever new phenomena' that opens up when we inquire 'back into what was ... taken for granted'. As noted, one key 'taken-for-granted' presupposition that has been challenged by the continental tradition since Husserl concerns the status he assigns to the subject as part of this process. A different problem gets foregrounded by his reference to 'infinity', however: in enacting a sense of the transcendental, how do we maintain relevance, and how do we block the theoretical tendencies of transcendental approaches towards infinite regress?

This returns us to the problem of relevance raised at the beginning of this chapter. In principle, this problem is persistent for transcendental approaches because of the seeming 'infinity' of conditions they open up. The claim of this chapter, however, is that it can be met *in fact* through a sense of imaginative variation as a practice involving a situated and enacted sense of the transcendental. To draw this out, let me conclude this part with a short series of further variations involving three important figures in the continental tradition: Husserl, Heidegger and Deleuze.

Further into section 35 of *Ideas*, Husserl develops the example of the 'white paper' with which we began this chapter:

> The apprehension [of the paper] is a singling out, every perceived object having a background in experience. Around and about the paper lie books, pencils, ink-well, and so forth, and these in a certain sense are also 'perceived', perceptually there ...; but whilst I was turned towards the paper there was no turning in their direction. ... They appeared and yet were not singled out, were not posited on their own account. Every perception of a thing has such a zone of *background intuitions*. (Husserl 2002: 117)

Husserl's theoretical framework here concerns the phenomenology of what is literally perceived. His concern is to draw out what he takes to be an invariant feature of the structure of perception: that it always takes place in terms of a foreground and background, such that whatever is explicitly perceived will be surrounded by a 'zone' or 'field' of what is not explicitly perceived or 'posited on its own account' (see also Ihde 2012: 35–44; Gurwitsch 2010: 1–10).

But what are the limits of this 'background'? Insofar as we remain within Husserl's framework, this question does not seem to arise: attending simply to what is given, it seems apparent *that* a given instance of sense perception has limits, independent of further questions concerning *why* and *how* it has them. Insofar as we can shift philosophical frameworks for considering this example, however, and insofar as the term 'background' invites figurative extension, the question can be explored in different ways.

An obvious reference point for taking things further here is the sense of the 'worldhood of the world' developed by Heidegger in *Being and Time*. Considered in this context, Husserl's example shows up differently: whereas Husserl wants us to focus on the background of his paper in a literal sense, describing mundane objects accessible to vision ('books, pencils, ink-well'), Heidegger invites us to consider conditions that are 'background' for Husserl's situation in a more wide-ranging sense.

Instead of characterizing Husserl's examples as primarily objects of vision, Heidegger's approach invites us to consider them as tools forming a network of relations which make up the sense of the 'worldhood of the world' as a meaningfully connected whole (Heidegger 2005: 97). To take up his paper, books, pencils and ink-well as examples of (actually or potentially) perceived 'present-at-hand' objects, on this account, Husserl must foreground them in a way that hides a more fundamental sense in which they are normally part of the 'background' of his situation: as tools or 'equipment' that he depends on to work, as 'ready-to-hand' (*Zuhanden*) (Heidegger 2005: 97).

But Heidegger is by no means the only thinker in the continental tradition whose approach allows us to further explore the limits of Husserl's 'background'. Consider, for instance, the following remarks from Deleuze on the artist Francis Bacon and the process of beginning to paint:

> It is a mistake to think that the painter works on a white surface. ... The painter has many things in his head, or around him, or in his studio. Now everything he has in his head or around him is already in the canvas, more or less virtually, more or less actually, before he begins his work. ... [It is] all present in the canvas. (Deleuze 2005b: 61–2)

On one level, it would be easy to write off the connections between Husserl's white paper and Bacon's canvas suggested here as naïve or superficial. According to one version of this criticism, citing this passage in this context may seem to imply that acts of writing and painting must both involve solitary creative 'geniuses'. What this would overlook, however, are the important connections between Husserl's white paper and Bacon's canvas that occur, not on the level of traditional aesthetic categories, but in terms of the status of their practices as situated, enacted and *technologically mediated*.

Deleuze's example is not simply an aesthetic reflection on painting in a traditional sense. Instead, it is better framed as a provocation to go as far as we can in inquiring into the virtual 'background' that constitutes the 'actuality' of a painter's situation. What the painter has 'in his head or around him', in this sense, is literally anything that can be imagined as a background condition or influence on the situation of their practice.[19] Like Husserl, the painter approaches what is ostensibly a 'blank page'; in fact, however, they are negotiating a highly conditioned space that implicates a range of conditions that take on 'more or less' relevance, and that shift in relevance contextually, over time.

How, when faced with this extent of conditioning, do situated and enacted practices respond to the problem of relevance raised at the beginning of this chapter? On this, Deleuze offers the following counterintuitive and suggestive response:

> The painter does not have to cover a blank surface, but rather would have to empty it. ... What we have to define are all [the] 'givens' that are on the canvas before the painter's work begins, and determine, among these ..., which are an obstacle [and] which are a help. ... A whole category of things that could be termed *'clichés'* already fill the canvas, before the beginning. ... *Clichés* are always already on the canvas. (Deleuze 2005b: 61–2)

On Deleuze's account, a *cliché* is an engrained condition placed upon a practice that inhibits the emergence of new and different responses to problems with which the practice is engaged (the problems of 'painting', for instance) (Deleuze 2004a). What is required to mitigate the effects of such *clichés* in the above example, in turn, is a sense of how creative practices involve acts of situated and technologically mediated imaginative variation. In this case, the act of variation involves virtually 'emptying' a canvas that is ostensibly 'clear', as a preliminary to working out which conditions are relevant to addressing the problems of painting. What this in situ enactment of variation works to block, in practice, then, is a theoretical tendency to see the practice of painting itself in terms of an overwhelming and inhibiting series of *clichés*, in favour of setting conditions for novel responses to problems raised by this practice.

In a general sense, the point to be emphasized here is this: while a sense of the transcendental can always slide towards an infinite regress of *'clichés'* in theory, it can always be channelled and enacted *in practice* through acts of imaginative variation as a way of introducing novelty into a context, and of addressing its problems of relevance.[20]

4 Re-imagining relevance (2)

Let me conclude this chapter by attempting to take Deleuze's sense of the *cliché* a little further. As we saw at the beginning of the previous part, one *cliché* affecting the practice of philosophy today is that approaches rooted in the 'continental' and 'analytic' traditions in Western philosophy can appear incompatible, without any way of meeting up to explore shared problems. Regrettably, this *cliché* also seems to affect the approach that Franssen and Koller advocate, despite the apparent openness we identified at the beginning of the previous part (on this, see Mitcham 2002). Writing as part of the team that curates the Stanford Encyclopaedia of Philosophy article on philosophy of technology, for instance, Franssen observes:

> During the last two centuries, when it gradually emerged as a discipline, philosophy of technology has been mostly concerned with the impact of technology on society and culture, rather than with technology itself. ... Only recently a branch of the philosophy of technology has developed that is concerned with technology itself and that aims to understand both the practice of designing and creating artifacts (in a wide sense, including artificial processes and systems) and the nature of the things so created. This latter branch of the philosophy of technology seeks continuity with the philosophy of science and with several other fields in the analytic tradition in modern philosophy, such as the philosophy of action and decision-making, rather than with social sciences and the humanities. (Franssen et al. 2013)

Unlike the passage from Franssen and Koller cited previously, which appeared open on how to extend philosophy of technology's scope and method, the above passage sets a number of more or less arbitrary restrictions on such a project. To draw this out, consider its presuppositions on three related matters: 1) 'technology itself'; 2) 'the practice of designing and creating artifacts'; and 3) the disjunction that Franssen et al. presuppose between 'the analytic tradition in modern philosophy' and 'social sciences and the humanities'.

Franssen et al. (2013) refer to 'technology itself' twice in the above passage, but in neither instance is the reference as rhetorically neutral as it might appear. Instead, it is invoked to assert the credentials of the emerging analytic approach to philosophy of technology that they favour against those of a more traditional (and, we might add, 'continental') approach that has been concerned with 'the impact of technology on society and culture'

(Franssen et al. 2013). Here, Franssen et al. seem to presuppose that we must either be doing positivistic philosophy of technology or some form of 'constructivist' philosophy of technology.

If we adopted opposed constructivist presuppositions, we might be inclined to respond with this question: what could 'technology itself' possibly be, in isolation from 'the impact of technology on society and culture'? This question itself becomes problematic, however, if all it does is return us to a presupposed exclusive either/or between positivism and constructivism in philosophy of technology. This is because turning exclusively in either one of these directions counts against the potential for the scope and method of philosophy of technology to expand in both directions at once: in an empirical direction focused on expanding and nuancing our sense of what can count as an object of inquiry, and in a transcendental direction focused on a wide-ranging sense of the conditions for the empirical.

This leads us to Franssen et al.'s consideration of the practice of 'designing and creating artifacts', in a 'wide sense'. Rightly, they emphasize that preconceptions on what is involved in the practice of design might be expanded so as to include artefacts that may seem counterintuitive when considered in terms of a common-sense view of what counts as 'Technology' (as encapsulated by their reference to 'artificial processes and systems'). Franssen et al. do not comment, however, on whether this consideration should be expanded to include 'exceptional technologies' such as the blank page, with all the shifting senses it implies.

In view of the misgivings Franssen et al. express against approaches focused on 'society and culture', it is doubtful they would accept such a focus as part of their definition of 'technology itself'. The reasons for this exclusion, however, appear arbitrary and constraining. This is because they count against the project of extending the scope and method of philosophy of technology so as to consider artefacts that run counter to our common-sense view of technology, and yet that, like 'blank pages', are also historically ubiquitous throughout diverse societies and cultures (and that have had manifold different material and normative impacts across these societies and cultures).

What also gets overlooked by Franssen et al.'s exclusion in this sense, and in a way that is arguably more important for the concerns of their own project, is a consideration of the propensity that any technology whatsoever has, in principle, to get taken up in diverse figurative ways. As I have attempted to show in this chapter, this propensity can be shown in diverse ways in the case of the blank page. Crucially, however, it is a propensity that can and does affect artefacts like the 'artificial processes and systems' to which Franssen et al. refer.[21] Franssen et al. seem to imply that this propensity is extrinsic to 'technology itself', and to do with 'society and culture'. In contrast, the suggestion of this chapter is that any approach aiming to be more empirically focused on 'technology itself' in a 'wide sense' has to be more open to seeing it as a condition affecting any artefact whatsoever, intrinsically.

This leads to a last point: Franssen et al.'s disjunction between 'the analytic tradition in modern philosophy' and the 'social sciences and the humanities'. There is no reason to presuppose, as Franssen et al. seem to, that this disjunction should be an exclusive one. Instead, it should be inclusive – given a particular technological artefact, problem or practice, 'X', we should be permitted and encouraged to imagine inquiries into 'X' that draw on either cutting-edge analytic philosophy, on work in the 'social sciences and humanities', on some novel combination of both or on other traditions and fields of thought. Otherwise, we run the risk of ruling out, in principle, imaginative crossovers that may be beneficial in fact, both for theory and for practical work in fields like design and engineering, whether in intended or in unintended ways.

In this chapter, I have suggested the 'blank page' as a case study of an exceptional technology that can be shown to challenge our sense of what 'technology itself' is, by virtue of the shifting literal and figurative senses it implies. One way of recognizing and drawing out the consequences of this, I have suggested, is through a practice of imaginative variation that comprises a situated and enacted sense of the transcendental, and that blocks the theoretical tendency of this sense towards an infinite regress of conditions. While retaining an empirical focus on the specificity of artefacts, practices and problems, this practice seeks to open and vary the range of conditions in relation to which these practices can be considered. It does so with a view to drawing out and generating novel consequences, both for the process of designing and creating artefacts and for philosophical inquiries into them.

CHAPTER THREE

Embodiment conditions

Through a focus on the theme of the blank page, the previous chapter sought to show how even an ostensibly trivial and everyday artefact can turn out to be an exceptional technology. The aim of this chapter runs in tandem: it seeks to show how a sense of the transcendental as an approach to argument or method, as outlined in Chapter 1, can already be detected across recent considerations of technology that draw on the continental philosophical tradition. The premise for proceeding in this way is that a developed and dynamic sense of the transcendental is methodologically appropriate to help us engage with exceptional technologies: while a sense of the transcendental focuses on conditions, a focus on exceptional technologies allows for dynamic examples that draw out conditions that otherwise might go unnoticed.

The chapter again focuses on a specific theme: embodiment. Drawing on figures such as Nietzsche, Bergson, Husserl, Merleau-Ponty, Deleuze and Haraway, a great deal of recent work in philosophy of technology and media theory emphasizes embodiment as a key theme for understanding technologies (Hansen 2006; Wegenstein 2010). Among those exemplifying this approach are Hubert Dreyfus, N. Katherine Hayles, Mark B. N. Hansen, Mark Poster, Brian Massumi, Wendy Chun, Catherine Malabou and Alva Noë. These figures often come from different disciplinary backgrounds, and their work often has different emphases. Without seeking to overlook these differences, the aim of this chapter is to emphasize a broad form of argument that such approaches can be viewed as sharing: characteristically, a key aspect of these approaches involves critical intervention against the tendency of the philosophical tradition to conceptualize technologies in ways that are abstract, instrumentalist or dualist, and that overlook the implied complexities of embodiment conditions.

The argument of this chapter is that such interventions demonstrate a powerful sense of the transcendental, and that making this explicit has the potential to focus, consolidate and extend a continental approach to the philosophy of technology. Of these claims, I take the first to be uncontroversial, but recognize the second to be problematic if not approached appropriately. This has to do with how differences between approaches to embodiment get handled. On the one hand, it is relatively straightforward to show how such approaches share a sense of the transcendental: given a prevailing conception of technology they take to be inadequate, they critique this by highlighting embodiment conditions it overlooks. The problem, however, is that highlighting such a shared form of argument may seem to do violence to the specificity of the approaches in which it is detected. Put simply, is highlighting a shared form of argument too schematic and formal a matter, and does it do violence to the plurality of different embodiment conditions to which approaches such as those highlighted above are committed?[1]

The contention of this chapter is that, rather than doing violence, drawing out a shared sense of the transcendental across different approaches can in fact strengthen and nuance our approach to the specificity of embodiment conditions. It develops this over four parts.

Part 1 conducts a critical reading of Dreyfus's *On the Internet*. It argues that while it would be easy to write this book off as naïve, anachronistic or insufficiently in depth, doing so would overlook the sense in which it provides an instructive example of advantages and difficulties facing approaches to philosophy of technology that are transcendentally focused on embodiment conditions. Against the temptation to write *On the Internet* off, I argue that it in fact exhibits an implicit sense of the internet as an exceptional technology, as well as a transcendental approach that has the capacity to draw this out, but that both these matters are insufficiently developed by Dreyfus.

Part 2 engages approaches to embodiment conditions emerging from recent media theory. The aim here is to expand the consideration beyond the restricted scope of Dreyfus's focus on the internet, and to outline the dynamics of a more expansive and reflexive transcendental approach. Through a focus on the work of N. Katherine Hayles and Mark B. N. Hansen, I argue that the lessons of work on embodiment in recent media theory need not be restricted to digital and internet-enabled new media, or to what Hansen calls 'twenty-first-century media'.[2] Instead, I argue that the approaches of Hayles, Hansen and other related thinkers comprise a body of transcendentally focused work that has the potential to be instructive for philosophical reflections on technology more broadly.

Part 3 considers how this work relates to an important area in recent philosophy of mind and cognitive science: '4e' approaches that situate cognition as 'embodied', 'embedded' and 'enacted' within an environment, and capable of 'extension' through technologies (Clark 1996, 2011; Rowlands 2010). The argument of this part is that work in this area exhibits a strong sense of the transcendental, and that, by virtue of this, it offers a refined taxonomy for complementing approaches to philosophy of technology and media theory that are focused on embodiment conditions.

Part 4 concludes by considering crossover potentials between the three main areas of work discussed in this chapter. The key claim of this part is that the sense of the transcendental emerging from work focused on embodiment conditions in philosophy of technology, media theory and 4e is compatible with an approach focused on exceptional technologies.

1 *On the Internet*

Hubert Dreyfus's short book *On the Internet* was published in 2001. Building on his more extensive books *What Computers Can't Do* (1992) and *Being-in-the-World* (1991), it develops a critique of forms of internet-centred transhumanism that emerged in the mid-to-late 1990s, such as 'Extropianism' and the Ray Kurzweil-inspired 'Singularity' movement (Dreyfus 2001: 4). Dreyfus argues that such approaches are extreme expressions of the tendency of the internet to make human beings overlook embodiment conditions as necessary for their being-in-the-world, and his critique has two main strands, both rooted in the continental philosophical tradition: a phenomenological strand drawing on Merleau-Ponty and Heidegger, and an existentialist strand drawing on some of the more polemical works of Nietzsche and Kierkegaard.

It would be easy to write off *On the Internet* as not deserving much attention today. For one thing, the forms of transhumanism that are its focus have so often been the objects of critique since that Dreyfus's approach may seem clichéd to the contemporary reader, and perhaps undue in the emphasis it gives these targets (Wegenstein 2010: 26–7; Bostrom 2005). Similarly, it would be easy to write off the phenomenological and existentialist strands of Dreyfus's approach as perhaps too traditionally 'philosophical' or subject-centred to deal with the empirical complexities of the contemporary internet, whether sociological, economic or technological.[3] At the limit, it might seem justifiable to reject Dreyfus's text as an anachronistic example of some of the worst tendencies of what was discussed in the previous chapter in terms of 'humanities philosophy of technology' (Franssen et al. 2013).

What such a reading would overlook, however, is that *On the Internet* provides an instructive example of some of the key advantages and difficulties involved in pursuing a transcendentally focused approach to embodiment conditions in philosophy of technology. My aim for this part is to conduct this overlooked reading. I will first consider how Dreyfus defines the internet and the human body, before then highlighting the sense in which his argument is 'transcendental'.

Here is how Dreyfus defines the internet at the beginning of his book:

> The Internet is not just a new technological innovation; it is a new type of technological innovation; one that brings out the very essence of technology. ... We have come to realize that the Net is too gigantic and protean for us to think of it as a device for satisfying *any* specific need. ... If the essence of technology is to make *everything* easily accessible and optimisable, then the Internet is the perfect technological device. It is the culmination of the same tendency to make everything as flexible as possible that has led us to digitalize and interconnect as much of reality as we can. What the Web will allow us to do is literally unlimited. This pure flexibility naturally leads people to vie for outrageous predictions as to what the Net will become. (2001: 1–2, Original emphasis)

Two aspects of this definition are immediately striking. First, it situates Dreyfus's approach as markedly Heideggerian. This is clear from his remark that the internet 'bring[s] out the very essence of technology', and that this 'essence' consists of the tendency 'to make *everything* easily accessible and optimisable'. These remarks are of a piece with Heidegger's famous claim that the essence of technology consists in the tendency to 'enframe' reality as 'standing reserve' ('*Bestand*') (Heidegger 1977). Second, Dreyfus's definition involves at least two problematic equivocations: Whereas he uses the terms 'net' and 'web' interchangeably, these refer to two distinct entities in a technical sense;[4] similarly, Dreyfus's description of the internet as a 'device' seems inappropriate given its technical status, not as a device, but as a network (see Galloway and Thacker 2007).

These points may seem simply to confirm Dreyfus's text as insufficiently empirical, and to strengthen the case for rejecting it. Doing so, however, would overlook something more important in his attempts at definition: the sense in which they implicitly recognize the internet as an exception to established forms of common sense on what constitutes a 'Technology'.

This recognition is, I think, what Dreyfus is grappling to make sense of in his characterizations of the internet as 'gigantic and protean' and 'a new type of technological innovation'. It is, moreover, also tenable to view it as a condition for the equivocations just identified: what makes these possible is that the ontology of the internet is not a settled or transparent matter, and that its status as a network problematizes common-sense conceptions of technology that are focused on apparently discrete and human scale 'devices', whether of the order of Heidegger's hammer, Galileo's telescope, a jet engine or an fMRI scanner.[5] The least that can be said about these complexities, I think, is that they do not merely affect Dreyfus's text; rather, they affect how contemporary human thought and natural languages in general struggle to conceptualize an entity like the internet.[6] In struggling to define what the internet is, then, it is arguable that Dreyfus is merely exhibiting some of these issues in a concentrated way.

Now consider how Dreyfus defines the human body:

> According to the most extreme Net enthusiasts, the long-range promise of the Net is that each of us will somehow be able to transcend the limits imposed on us by our body. ... By our body, such visionaries seem to mean not only our physical body with its front and back, arms and legs, and ability to move around in the world, but also our moods that make things matter to us, our location in a particular context where we have to cope with things and people, and the many ways we are exposed to disappointment and failure as well as to injury and death. In short, by embodiment, they include all aspects of our finitude and vulnerability. In the rest of this book, I will understand the body in these broad terms. (2001: 4)

What is immediately striking about this definition is that it resembles less the body as conceived of by transhumanists (see, for instance, Kurzweil 1999), and more the body as conceived of by Dreyfus's own phenomenological and existentialist influences.[7]

Dreyfus's implicit justification for this rhetorical sleight of hand concerns the complexity of phenomenological and existential conditions concerning physical location, movement, mood, finitude and vulnerability that he takes 'the body' to imply, in all circumstances. This sense of the body is indicative of Dreyfus's own phenomenological and existentialist commitments. However, the way he makes his point might easily be criticized for making a straw man out of transhumanism, and for begging the question: Dreyfus does not indicate who his targets are in the above passage, referring instead to an unspecified set of 'visionaries' and 'Net enthusiasts', and he can only criticize this group for overlooking the types of conditions that phenomenology and existentialism classically treat of by presupposing that such conditions are what define the human body.

Again, however, it would be a mistake to take these points as sufficient to dispense with what Dreyfus has to tell us about embodiment. This is because stopping short like this would cause us to overlook a more important issue concerning 'the body' that goes on in Dreyfus's definition. Put simply, what is problematic about Dreyfus's approach is not the fact that he draws our attention to the phenomenological and existential conditions of embodiment to which he is committed; rather, it is the fact that he conflates a notion of 'the body' per se with these conditions. In fact, charges of rhetorical sleight of hand might be escaped simply by shifting the terms of approach: all that may be required is to characterize things, not as an attempt to define 'the body' once and for all and across all contexts, but rather as part of an ongoing inquiry into a plurality of *embodiment conditions*.[8]

We have now looked at how Dreyfus defines both the internet and the human body. In both cases, we found problems with his approach, but also implicit potential that focusing only on the problems might cause us to overlook. Now consider how the argument of *On the Internet* relates these two terms. Here is Dreyfus's summary:

> We should remain open to the possibility that, when we enter cyberspace and leave behind our animal-shaped, emotional, intuitive, situated, vulnerable, embodied selves, and thereby gain a remarkable new freedom never before available to human beings, we might, at the same time, necessarily lose some of our crucial capacities: our ability to make sense of things so as to distinguish the relevant from the irrelevant, our sense of the seriousness of success and failure that is necessary for learning, and our need to get a maximum grip on the world that gives us our sense of the reality of things. Furthermore, we would be tempted to avoid the risk of genuine commitment, and so lose our sense of what gives meaning to our lives. ... I hope to show that, if our body goes, so does relevance, skill, reality, and meaning. If that is the trade-off, the prospect of living our lives in and through the Web may not be so attractive after all. (2001: 6–7)

Dreyfus develops this argument over four chapters. In Chapter 1, he argues that human embodiment is a necessary condition for solving problems of relevance and retrieval that are posed by the scale of information available online (2001: 8–26). In Chapter 2, he argues that embodied involvement and risk are necessary for skill acquisition and expertise (2001: 27–49). In Chapter 3, he argues that a host of what Alva Noë has called 'varieties of presence' are necessary for the more reduced representations of the self that occur online in 'telepresence' (Noë 2012; Dreyfus 2001: 50–72). In Chapter 4, he argues that embodied emotions and moods are necessary to give meaning to a shared human lifeworld (2001: 73–89).

The point here is that the argument of *On the Internet* is resolutely 'transcendental', both as a whole and in each of its parts. At each stage, Dreyfus takes particular problems posed by the internet and argues that embodiment conditions are necessary but overlooked for making sense of them. This noted, however, there is a case for further developing the sense of the transcendental that his argument exhibits.

As we saw, Dreyfus's definition of the internet is problematically equivocal, and his definition of the body conflates it with a particular set of phenomenological and existential conditions. A way of tempering this would consist in going further with the sense of the transcendental that Dreyfus's text already exhibits: in doing so, we can make explicit a sense of the internet as an 'exceptional technology' that is already implicit in Dreyfus's approach, and, correlatively, we can begin to inquire into a plurality of embodiment conditions that are overlooked by his text.

By claiming that the internet is an 'exceptional technology', I mean that it is an exception to received forms of common sense on what empirically constitutes a technology. Stated differently: the internet is, by virtue of its scale, protean nature, status as a network, as well as the plurality of technological, sociological, political and economic concerns it implicates, a technology that is '*para*-doxical'. That is: the internet is a technology that runs counter to established common beliefs ('*doxas*') concerning the status of technologies as, for instance, more obviously fitted to the scale of human perception, manipulable and tangible, mechanical, recyclable, or as not obviously 'networked' (see also Smith 2015).[9]

As noted above, it is tenable to argue that *On the Internet* implicitly recognizes this status of the internet, and that this is a condition for Dreyfus's difficulties in defining it. Going further, we can, I think, characterize it as a condition for Dreyfus's endorsement of a Heideggerian approach. Having implicitly recognized the exceptional character of the internet, Dreyfus makes an unfortunate move: instead of going further in a transcendental sense and developing an approach more suited to engaging the specificities of the case of the internet, he has recourse to a philosophy of technology with which he is familiar, and finds it in a Heideggerian register concerning the 'essence' and 'tendency' of technology. On the account developed in this book, this is a radically insufficient move that tends towards reifying 'Technology' as a whole, and that forsakes an opportunity: to engage the internet, not as yet another exemplar of some reified essence, but as an 'exceptional technology' – that is, as a technology that, in a series of very specific and complex ways, exceeds and challenges received conceptions of what constitutes technologies (including the received Heideggerian conception).

As Dreyfus recognizes, a key way the internet challenges such received conceptions concerns the necessity for taking embodiment conditions into account. As we saw, however, his definition of the body conflates it with a particular set of conditions. A more thoroughgoing approach would, I suggest, involve going further with the sense of the transcendental that Dreyfus's move to consider the body implies. This would allow us to engage, not merely with phenomenological and existential conditions, but, for instance, with historical, genetic, economic and political ones as well.

This move would involve recognizing that any consideration of embodiment must be open-ended, cross-disciplinary, and that it must be an inquiry into a plurality of conditions. Rather than running the risk of elevating a particular conception of 'the body' as a norm, or of mistaking a part of the body for the whole of its conditions (the brain as conceived of by classical cognitive science, for instance), this approach would strive to be more methodologically responsive to issues concerning, for instance, the specificity of body parts and the relations between them, as well as issues concerning gender, ethnicity, situation, age, disability, body enhancement, and issues concerning the method of appraisal used (neuroscience as opposed to evolutionary biology, for instance).[10]

2 A developing body of work

I have just argued that *On the Internet* exhibits a sense of the transcendental that has much greater potential than is developed in the text. Even if this is granted, however, there might still be a case for writing off Dreyfus's approach as an isolated example. Similarly, even if Dreyfus's argument has interesting (and contestable) things to say about embodiment conditions, it might be doubted that his approach has much to tell us about other technologies, beyond the case of the internet. The aim of this part is to mitigate these suspicions by detecting signs of a more thoroughgoing sense of the transcendental in other work on embodiment conditions that is influenced by the continental philosophical tradition. The argument is that a powerful body of such work already exists in recent approaches to media theory, and that the findings of this work have the capacity to extend beyond a focus on digital and internet-enabled 'new media'.

Consider, for instance, these remarks from the conclusion to N. Katherine Hayles's influential book *How We Became Posthuman*:

> Human being is first of all embodied being, and the complexities of this embodiment mean that human awareness unfolds in ways very different from those of intelligence embodied in cybernetic machines. ... The body is the net result of thousands of years of sedimented evolutionary history, and it is naïve to think that this history does not affect human behaviours at every level of thought and action. (1999: 284)

When Hayles asserts that 'human being is first of all embodied being', she means that embodiment conditions are necessary for understanding human being-in-the-world with technologies, and that they cannot be overlooked for either theoretical or practical purposes. Like Dreyfus's argument, then, Hayles's approach demonstrates a strong sense of the transcendental as an approach to argument or method.

Now consider two aspects of Hayles's summary that already nuance this sense and take it further than Dreyfus's approach. First, Dreyfus consistently entertained the possibility that disembodiment through the internet may, for the purposes of argument, 'somehow' be achievable. A crucial feature of Hayles's work, in contrast, is to emphasize that information technologies also necessarily imply embodiment conditions. This nuances and expands attention to the *materiality* of such conditions: for Hayles, information never 'loses its body' (1999: 2), and even the intelligence of 'cybernetic machines' has embodiment conditions; it is just that these are materially 'very different' from those affecting human awareness (see also Hayles 2012: 3, 17). Second, it should also be noted that Hayles goes further than Dreyfus in drawing attention to the historical and evolved character of embodiment conditions: whereas Dreyfus's focus on classically phenomenological and existential conditions runs the risk of appearing too traditionally focused on the (conscious) subject, and perhaps conservative in its conception of the body thereby, Hayles's emphasis on historically changing material conditions allows her to highlight the body's capacity to 'become', whether in ways that are conscious, volitional, and arrived at in the short-term, or more unconscious, environmental and longitudinal (see also Hayles 2012: 85–121).

These two points converge in Hayles's concept of the 'posthuman'. Insofar as human beings are entangled within environmental niches where embodiment conditions on intelligence are recognized not to be limited to the human, and where the interaction and becoming of bodies is conceived of in more processual ways, we have, on Hayles's account, already entered a 'posthuman' era. Indeed, one of the key stakes of Hayles's work consists in thinking through what the future implications of this, our present 'posthuman condition', might be:

> If my nightmare is a culture inhabited by posthumans who regard their bodies as fashion accessories rather than the ground of being, my dream is a version of the posthuman that embraces the possibilities of information technologies without being seduced by fantasies of unlimited power and disembodied immortality, that recognizes and celebrates finitude as a condition of human being, and that understands human life is embedded in a material world of great complexity, one on which we depend for our continued survival. (1999: 5)

On the one hand, there are strong parallels with Dreyfus here, especially in the emphasis Hayles places on finitude and the body as the 'ground of being'. That said, Hayles is much keener than Dreyfus to point out what she sees as the properly ideological character of human 'fantasies' of disembodiment, as well as potentials for what she calls the 'coevolution' of human and technological bodies, by virtue of participation in a shared 'material world' (see 2012: 90–1).

There would, on this basis, be a localized case here for viewing Hayles's approach as a strong supplement and corrective to the sense of embodiment developed by Dreyfus in *On the Internet*. Stronger than this, however, there is a much broader methodological case for viewing her approach as extending and nuancing of our sense of the transcendental. Put simply, Hayles takes the inquiry into embodiment conditions further than Dreyfus. This is because she goes beyond the focus on classically phenomenological and existential conditions exhibited by his approach, in favour of a more thoroughgoing focus on the historical and evolutionary conditions in which these conditions are nested, and from which they emerge (see also Varela et al. 1991).

In some of her more recent work, Hayles has framed issues concerning the coevolution of humans and technologies in terms of 'technogenesis' (2012).

This concept is drawn from the work of the French philosopher Gilbert Simondon, and Hayles's use of it opens important points of connection and contrast between her work and that of other media theorists, including, notably, Mark B. N. Hansen.

In *New Philosophy for New Media*, Hansen draws on Bergson's approach to the body to engage with work from new media artists, including Alba d'Urbano, Douglas Gordon and Bill Viola. From the outset, however, he is at pains to emphasize that he is after something more than a piecemeal collection of examples:

> Rather than a survey of new media art, my study aims to theorize the correlation of new media and embodiment. Toward this end, I have found it most useful to focus on works by new media artists that foreground the shift from the visual to the affective registers and thereby invest in the multimedia basis of vision itself. In this sense, my decision is above all a strategic one: if I can prove my thesis (that the digital image demarcates an embodied processing of information) in the case of the most disembodied register of aesthetic experience, I will, in effect, have proven it for the more embodied registers (e.g. touch and hearing). ... Moreover this strategic decision resonates with the interests of contemporary artists themselves: even those artists not directly invested in these embodied registers can be said to pursue an aesthetic program aimed first and foremost at dismantling the supposed purity of vision and exposing its dirty, embodied underside. (2004: 11–12)

We can, on the basis of these remarks, be quite explicit about the 'something more' Hansen is seeking: the 'new philosophy for new media' he is after is a philosophy of embodiment that exhibits a thoroughgoing sense of the transcendental.

Consider the presuppositions underpinning Hansen's stated intentions. First, his 'correlation' between new media and embodiment involves a strong and dynamic sense of the transcendental, because it presupposes that new media and embodiment reciprocally condition one another. Second, his 'strategic decision' to foreground artistic work that takes the shift from 'the visual to the affective register' as a theme presupposes that the visual has the affective as a condition of possibility, or, as Hansen puts it, its 'multimedia basis' (this, in turn, is what is behind his central claim that vision is 'the most [apparently] disembodied register', in contrast to other senses like touch and hearing). Third, Hansen's recourse to contemporary artists in his final clause presupposes that art and philosophy are involved in a shared transcendental project to investigate vision's embodied conditions.[11]

In turn, these presuppositions inform what Hansen takes to be 'new' about new media:

> Beneath any concrete 'technical' image or frame lies what I shall call the *framing function* of the human body. ... This ... correlates directly

with the so-called digital revolution. If the embodied basis of the image is something we can grasp only now, that is because the so-called digital image explodes the stability of the technical image in any of its concrete theorizations. Following its digitization, the image can no longer be understood as a fixed and objective viewpoint on 'reality' – whether it be theorized as frame, window, or mirror – since it is now defined precisely through its almost complete flexibility and addressability, its numerical basis, and its constitutive 'virtuality'. (2000: 8, Original emphasis)

Framed in the terms of this chapter, what is new about digital images and the new media technologies that make them possible on Hansen's account is that they challenge established forms of common sense (whether more or less everyday or theoretical) on what constitutes an 'image', a 'medium' or a 'Technology'. This is what is implied in Hansen's above remark on the 'explosion' implied by the so-called digital image. What is new about 'new media', then, on Hansen's account, is that they show up as exceptional in relation to our established sense of things, and this, in turn, is why he thinks a new media philosophy, exhibiting a more thoroughgoing sense of the transcendental, is required to make sense of new media.[12]

But doesn't this emphasis on 'new media', evident in the work of Hayles and Hansen alike, pose a significant problem?[13] Doesn't it imply that any lessons drawn from their work will be restricted to new media and that they have little to tell us about technologies in any broader sense? In what remains of this part, I will argue, on the contrary, that the sense of the transcendental demonstrated by their work, and that of related new media theorists, is instructive for how a continentally informed philosophy of technology might seek to engage the specificity of technologies and their conditions in general.

While Hayles and Hansen emphasize new media, the subject matter of their work is not drawn exclusively from this domain. On the contrary, both thinkers include a focus on 'old media' technologies, and, in both cases, the consideration of 'old' and 'new' media alike is inflected in very particular ways by other disciplines (most obviously, literary theory in Hayles's case, and aesthetics in Hansen's). Rather than viewing Hayles and Hansen's work as offering a Whiggish account of history inevitably leading to 'new media', then, it is more accurate to view them as offering complex and contrasting 'genealogies' or 'archaeologies' of media, in a sense that resonates both with Foucault's sense of these terms, and with the contemporary 'media archaeology' movement (Foucault 2002; Parikka 2012). The point is that Hayles and Hansen do not simply take 'new media' as an all-consuming given that requires them to pay lip service to older technologies. Instead, their work is underpinned by recognition of the need to offer critical readings that look at the history of new media to emphasize the contingency of its emergence and present forms.[14]

What should also be emphasized is that Hayles and Hansen draw on a common set of continental philosophers who take technology as a theme, including, for instance, Merleau-Ponty and Simondon (see, e.g. Hayles 2012: 87–90; Hansen 2006: 82–94). On a very crude reading, it might be suspected that Hayles and Hansen take these approaches to have covered 'old media' in depth, and that they take this as warrant to read the history of philosophy in a way that cherry-picks aspects of the older approaches.[15] What this would overlook, however, are more profound methodological affinities that are common to the work of Hayles and Hansen, and to that of thinkers like Merleau-Ponty and Simondon alike. As Hansen puts it in *Bodies in Code*:

> Merleau-Ponty's phenomenology of embodiment is, from the beginning, a philosophy of embodied technics. ... Accordingly, one of our pressing tasks here will be to think this 'originary' technics as it might have been (but was not) developed by Merleau-Ponty, to think this technics beginning from but moving well beyond Merleau-Ponty's limited conception of prosthetics as the extension of bodily habit. In doing so, however, we must never lose sight of the fact that Merleau-Ponty ... himself gives us the means. (2006: 39)

These remarks make it clear that Hansen conceives of the relation between his work and that of Merleau-Ponty in terms of shared affinities of transcendental method. By this, I mean is that he considers his work, alongside that of Merleau-Ponty, to be involved in a shared and ongoing project to inquire into the historically changing conditions that are implicated by technologies. As Hansen recognizes, commitment to this involves critiquing the limits of Merleau-Ponty's approach, by positing overlooked conditions for its possibility. This does not commit Hansen to condemning Merleau-Ponty's approach to history, however, nor to stripping it for parts. Instead, Hansen's aim is to build upon and catalyse the potentials of Merleau-Ponty's work for a historically different set of circumstances and technologies.

A similar sense of the transcendental is also evident in Hayles's work on Simondon:

> I propose that attention is an essential component of technical change (although undertheorized in Simondon's account), for it creates from a background of technical ensembles some aspect of their physical characteristics upon which to focus, thus bringing into existence a new materiality that then becomes the context for technological innovation. Attention is not, however, removed or apart from the technological changes it brings about. ... Technical beings and living beings are involved in continuous reciprocal causation in which both groups change together in coordinated and indeed synergistic ways. (2012: 103–4)

These remarks demonstrate affinities for a transcendental approach that, in a manner parallel to Hansen's engagement with Merleau-Ponty, Hayles presupposes her work to share with that of Simondon. When Hayles states that attention is an 'essential component' of technical change, then, she means that it is a condition that requires a dynamic sense of the transcendental to be investigated. Likewise, when she states that this component is 'undertheorized' by Simondon, she implies that Simondon's work is involved in this investigation, but that he does not take it far enough. Lastly, when Hayles states that 'technical' and 'living' beings 'are involved in continuous reciprocal causation', she is making explicit a condition that her work shares with Simondon, and that she thinks must act as a condition for the more thoroughgoing theorization of attention she is seeking.[16]

Focusing on this shared sense of the transcendental enables us to view Hayles and Hansen's work, not as historically free-floating, syncretic and 'cherry-picking', or exclusively concerned with 'new media'. Instead, it allows us to position their work as part of a developing body of continentally influenced philosophy of technology and media theory, a key dimension of which involves a transcendental approach to embodiment conditions. This enables us to view Hayles and Hansen's work as an important critical touchstone in two respects. First, it can be compared and contrasted with the work of other contemporary thinkers who demonstrate similar affinities in terms of method and argument (see, for instance, Stiegler (2010), Chun (2011), Poster (2001b, 2006), Galloway (2012), Liu (2010)). Second, it points towards a method for engaging, not just with new media or 'twenty-first-century media', but with the specificities of technologies in general.

This second point has crucial implications for the approach developed, not merely in this part, but in this book as a whole. To draw these implications out, consider the following problem concerning the 'new' that Hansen identifies in *New Philosophy for New Media*:

> What is it about new media that makes them 'new' ...? For almost every claim advanced in support of the 'newness' of new media, it seems that an exception can readily be found, some earlier cultural or artistic practice that already displays the specific characteristics under issue. This situation has tended to polarize the discourse on new media art between two (in my opinion) equally problematic positions: those who feel that new media have changed everything and those who remain sceptical that there is anything at all about new media that is, in the end, truly new. No study of new media art can afford to skirt this crucial issue. (2000: 21)

On the account developed in this part, neither Hayles nor Hansen can be accused of skirting this issue. On the contrary, their approaches explore a shared way beyond the problem (a veritable 'antinomy') Hansen identifies here, and, furthermore, they do so in a way that has implications beyond *New Philosophy for New Media*'s focus on 'new media art'.[17] The reason for this is that, by focusing on the conditions that new media technologies implicate, Hayles and Hansen's respective approaches provoke us to recognize the issue of new media's 'newness' as one that is as much 'transcendental' as it is 'empirical'.[18]

Empirically, new media are subject to many of the same conditions as 'old media'. As Hayles called attention to above, for instance, new media always remain subject to conditions of materiality and embodiment, and it would be wrong to see them as 'dematerialized' in any absolute sense (on this, see also Reading 2014). Focusing on only such conditions, however, might cause us to stress continuity at the expense of discontinuity, and to reduce new media to 'old media'. What this would cause us to overlook, in turn, are the ways new media are empirically different from 'old media'. As Hansen called attention to above, for instance, new media are networked in more obvious ways than old media, and they involve digital rather than analogical content (on this, see also Galloway and Thacker 2007; Gere 2008). The problem with focusing only on such differences, however, is that it might cause us to stress discontinuity at the expense of continuity, and to see new media as having no relation whatsoever to 'old media'.

One way to account for both the continuities and the discontinuities at work here is to recognize that the changes at stake do not occur within a settled conception of the 'empirical'. Rather, the issue of new media's 'newness' is also the kind of issue that affects the conditions under which the empirical gets recognized. On the account developed in this part, this is the kind of recognition that the work of both Hayles and Hansen aims at, and it is why both thinkers develop approaches that aim at a dynamic and thoroughgoing sense of the transcendental.

As we have seen, this is explicitly the sense in which new media are 'new' for Hansen: by showing up as exceptions to established forms of common sense on what empirically constitutes 'images', 'media' and 'technologies', new media change the conditions under which 'images', 'media' and 'technologies' are recognized, and what is required to make sense of this, on his account, is a renewed sense of the transcendental (a 'new philosophy for new media').

Proceeding in this way allows us to account for why new media can simultaneously seem to have changed 'everything' and 'nothing'. If we focus on change in a transcendental sense, we might be inclined to think that new media have 'changed everything', because we will stress that they have changed the conditions under which technologies are recognized. If, conversely, we focus on change in an empirical sense, we might be inclined to think that they have changed nothing, because we will stress that they are made of the same 'stuff' as older media.

The difficult balancing act when faced with this is to arrive at an approach that, rather than emphasizing the transcendental at the expense of the empirical, or vice versa, manages to conceptualize their correlation in as reflexive and responsive a way as possible. In *New Philosophy for New Media*, Hansen's emphasis on the correlation between embodiment and new media recognizes the need for such an approach, but arguably does not take it far enough. This is because, having recognized that new media change the transcendental conditions under which images, media and technologies show up, Hansen makes the strategic decision to empirically restrict his study to new media art.

Hansen's reasons for this move have to do with the importance he assigns to new media art as a mode of revelation.[19] It should be noted that at least one further strategy is possible, however. The lesson this alternative strategy would take from Hansen's above remarks is not that all technologies and media should be reduced to either 'old' or 'new'. Instead, it would focus on the fact that, as Hansen observes, for any given attempt to institute such a reductive sense of what constitutes media or technologies, 'an exception can readily be found'. The lesson this alternative strategy would take, then, is this: to engage with and learn from such *exceptional technologies*, we have to arrive at a sense of the transcendental that is sufficiently reflexive to be capable of engaging with the specificities of old and new media and technologies alike.

In this part, I aimed to show how a developing body of work on embodiment conditions in media theory may already exhibit some of the potentials of this alternative strategy. In the next part, I will consider how these potentials might be developed further, in relation to an area of contemporary philosophical work that is ostensibly very different.

3 Situating embodiment conditions: 4e

My aim for this part is to relate approaches looked at so far in this chapter to work emerging from what Mark Rowlands has called the '4e' context in philosophy of mind and cognitive science.[20] For reasons of space, my aim is not to offer an extensive survey of work in this area. Instead, it is to highlight how work emerging from the 4e context can be viewed as exhibiting a sense of the transcendental that complements the approaches to embodiment conditions looked at so far. For this reason, I hope that the move to consider 4e at this stage will not seem like too jarring a shift.

In his 2010 book *The New Science*, Rowlands offers the following instructive (but highly compressed) summary of work in the 4e context:

> [Cognitive science's] new way of thinking about the mind is inspired by, and organized around, not the brain but some combination of the ideas that mental processes are (1) *embodied*, (2) *embedded*, (3) *enacted*, and (4) *extended*. ... The idea that mental processes are *embodied* is [roughly] that they are partly constituted by ... [extraneural] bodily structures and processes. The idea that mental processes are *embedded* is ... the idea that [they] have been designed to function only in tandem with a certain environment that lies outside the brain of the subject. ... The idea

that mental processes are *enacted* is the idea that they are made up not just of neural processes but also of things that the organism does more generally – that they are constituted in part by the ways in which an organism acts on the world and the ways in which world, as a result, acts back on that organism. The idea that mental processes are *extended* is the idea that they are not located exclusively inside an organism's head but extend out, in various ways, into the organism's environment. (2010: 3, Original emphasis)

The first thing I want to stress in picking through this summary is simply that it exhibits a strong sense of the transcendental: given the perceived inadequacies of previous forms of brain-focused cognitive science, the 'new [4e] way of thinking' involves a shift to consider conditions for cognitive processes that are covered over by such a 'brainbound' or residually 'Cartesian' focus (Rowlands 2010: 2–3; see also Wheeler 2005; Clark 2011; Damasio 2003).

The next important thing to stress is just how multifarious this sense of the transcendental becomes in the 4e context. What is key is that while approaches working in this context take a focus on embodiment to be necessary for understanding cognition, they do not take it to be sufficient. What is important about 4e research in this sense is that it offers a context for developing a taxonomy that is attentive to issues concerning if and when we should seek to differentiate embodiment conditions from those concerning 'embeddedness', 'enaction' and 'extension'. The important upshot of this, in turn, is that work in the 4e context both has something in common with the approaches to embodiment from Dreyfus, Hayles and Hansen that we have looked at so far over the course of this chapter (namely, a strong sense of the transcendental), and that it offers something importantly different: a taxonomy that can act as an analytical tool for taking this further in multifarious ways.

The approach that Rowlands himself goes on to develop in *The New Science* is exemplary of this. Having sketched out what he sees as the general context of 4e work, Rowlands argues for a '2e' approach focused on conditions of embodiment and extension (2010: 85–106). His argument for this approach, which he calls the 'amalgamated mind', is based on the following double claim: wherever approaches focused on conditions of enaction and embeddedness turn out not to reduce to conditions of embodiment and extension, he argues, they turn out to be unproblematic for the principles of what he calls brain-focused 'Cartesian cognitive science' (2010: 19, 21).

What ultimately renders 'enacted' and 'embedded' approaches unproblematic for old style 'Cartesian' approaches, according to Rowlands, is their tendency towards the weak empirical claim that practices and environments merely 'supplement' or 'drive' cognitive processes. As he puts it:

The claim that cognitive processes are dependent, even essentially dependent, on wider bodily structures and circumstances does not, in any way, force us to reject the claim that cognitive processes occur exclusively inside the brain. (2010: 57; see also Clark 2011: 112; Rupert 2004: 393)

Whether or not we ultimately buy into Rowlands's arguments for his 2e approach, what is more important to note here is the form his arguments take: in contrast to the 'dependence' claim just outlined, Rowlands favours a stronger claim that environments, bodies, artefacts and practices can in some (but not necessarily all) circumstances *constitute* cognitive processes (2010: 13). In other words, not merely are there circumstances in which cognitive processes depend, in fact, on what Rowlands calls the 'extraneural'; rather, there are circumstances where the extraneural can, in principle, be part of the cognitive processes (see also Clark 2011: 114–16).

This claim in favour of a constitutive role for the 'extraneural' comprises both the novelty and the controversy of much work in the 4e context. What is significant to note here, in turn, is that it demonstrates a sense of the transcendental as an approach to argument or method. In Rowlands's own exposition, this sense is explicit and highly attuned. Consider, for instance, these remarks on the concept of sense itself:

> In its transcendental role, sense occupies a noneliminable position in any intentional act. Any attempt to make sense into an object – and hence empirical – will require a sense in virtue of which this transformation can be accomplished. Moreover, it is to sense in its noneliminable transcendental role that we must look if we want to understand the intentionality of thought – the directedness of thought towards its object. (2010: 172–3)

The use of a phenomenological term like 'intentionality' should not mislead us here. It does not indicate that Rowlands has fallen back into some form of Cartesianism. Rather, the key stake of his approach, along with that of others working in the 4e context, is to offer a different account of the conditions constituting cognition than is offered by the 'internalism' of Cartesian approaches.[21] Rather than viewing cognition as an intentional process directed outward from the mind or brain of the subject, then, Rowlands makes sense of it in terms of a wider and more fundamental process of 'world disclosure' that can, in principle, be constituted by embodied processes, and by technological artefacts and extended environmental structures:

> Intentional directedness is best understood in terms of the idea of world disclosure. ... World disclosure ... is entirely neutral over the nature and location of its vehicles. Sometimes they are neural operations, but sometimes they are processes taking place in the body, or even processes that extend into the world in the form of manipulation, exploitation, and transformation of environmental structures. (2010: 218)

The register of 'world disclosure' and 'revealing' Rowlands employs here might immediately make us think of Heidegger (1977). It would, however, be inappropriate to view his approach as a straightforward updating of Heidegger, for two main reasons. First, as a matter of historical and textual influence, Rowlands draws on other treatments of the transcendental to develop his approach, including those of Kant (2010: 165), Frege (2010: 170), Husserl (2010: 174–8), and Sartre (2010: 178–81). Second, in emerging from the 4e context, his sense of the transcendental is a product of a paradigm whose taxonomy, terminology and concerns, while sometimes overlapping with those of Heidegger, are importantly distinct in many other respects.[22]

Rowlands is one of the thinkers working in the 4e context who is most explicit on the theme of the transcendental.[23] This does not mean that his work is an outlier, however. On the contrary, a strong and multifarious sense of the transcendental can be detected right across work in the 4e context. To draw this out, consider Clark and Chalmers's 1998 article 'The Extended Mind', a piece that is canonical for subsequent 4e work. In it, Clark and Chalmers advocate a form of 'active externalism' based on a principle of parity between cognitive resources internal to the human organism and external environmental supports:

> In [certain] cases, the human organism is linked with an external entity in a two-way interaction, creating a *coupled system* that can be seen as a cognitive system in its own right. All the components in the system play an active causal role, and they jointly govern behaviour in the same sort of way that cognition usually does. If we remove the external component the system's behavioural competence will drop, just as it would if we removed part of its brain. Our thesis is that this sort of coupled process counts equally well as a cognitive process, whether or not it is wholly in the head. (2011: 222, Original emphasis)

The really crucial statement here for our purposes occurs in the penultimate sentence. Clark and Chalmers write: 'If we remove the external component the system's behavioural competence will drop, just as it would if we removed part of its brain.' Consider the constitutive ambiguity of the term 'drop'. To the extent that 'dropping' admits of degrees (as in 'the temperature dropped'), it is possible to read Clark and Chalmers's approach as tending towards the kind of enactive or embedded claim that is, according to the account developed by Rowlands above, empirically obvious: that 'external components' provide supports or 'scaffolding' for cognitive processes that can go on in a 'dropped' capacity, and that remain internal in all essential respects. On the other hand, complete cessation is a possible degree that 'dropping' admits of (as in 'he dropped out of University'). In this case, a much stronger claim for active externalism emerges that in fact forms the core of the extended mind thesis, and that is consistent with Rowlands's favoured

'2e' approach, as discussed above: the claim that external components can, in certain circumstances, act as necessary constitutive conditions for cognitive processes that could not go on in their absence.

Such ambiguities have acted as important catalysts for subsequent 4e work. In debates on the scope of extended mind thesis itself, interpretations have tended between models of cognition that emphasize functionalism (Wheeler 2012; Clark 2008) and more liberal models that emphasize enactivism (Gallagher and Crisafi 2009; Malafouris 2013). While enactivists typically want to extend the scope of the extended mind thesis further, some functionalists take enactivism to tend towards forms of 'vital materialism' that are too broad and explanatorily weak (Wheeler 2012: 6).

Beyond the terms of this particular debate, many other issues are at stake in 4e work, from discussions of canonical problems that recur throughout the literature, to fine-grained issues concerning how different taxonomies are to be drawn up to classify the conditions of cognition. For reasons of space, I cannot enter into the specifics of these issues here.[24] Instead, let me conclude this part by reemphasizing a general point that can be viewed as important for them all: quite simply, a condition for further debates and interpretative issues arising at all in the context of 4e work is a strong and multifarious sense of the transcendental.

This point would be highly dubious and controversial on the understanding of the term 'transcendental' as connoting some form of otherworldly realm of essences. When viewed in terms of the sense of the transcendental as a dynamic approach to argument or method that I have sought to develop over the course of this book, however, I take it to be relatively uncontroversial. To draw this out, consider Clark and Chalmers's discussion of 'dropping' again: questions of degree surrounding this are ambiguous and this has acted as a catalyst for further debates in the 4e context; what is not ambiguous across these debates, however, is that work in this context is focused on the conditions for the possibility of cognition, in a dynamic and expansive way that challenges the presuppositions of 'brainbound' Cartesian approaches. What is at issue in 4e debates, then, is not whether a sense of the transcendental as an approach to argument or method is appropriate to advance work in this context. Instead, this tends to be trivially accepted. What is at issue is how strong this sense of the transcendental should be, and which conditions it should be focused upon.[25]

4 Crossover potentials: Between philosophy of technology, media theory and 4e

Let me try to draw some of the stakes of the issue just identified into sharper focus. My aim in doing so is to highlight the ways in which considerable potentials for crossover exist between each of the main areas of work discussed in this chapter, and to show how these potentials are compatible with the approach focused on 'exceptional technologies' forwarded in this book.

In an article emerging from the liberal end of the spectrum in 4e debates, Shaun Gallagher and Anthony Crisafi write the following in favour of what Gallagher calls the 'socially extended mind' thesis (2013):

> There is no good reason, once we start along the path of the extended mind, to stop short of considering ... larger processes, such as ... processes involved in social, educational, and legal institutions, as cases of extended cognition. (2009: 51)

The claim here is that there is no reason, in principle, not to extend our consideration of the constitutive conditions for cognition beyond what Gallagher identifies as a restrictive set of 'typical examples ... rehearsed in the extended mind literature' (2013).[26] For Gallagher and Crisafi, this is something to be celebrated, and involves stretching the scope of the extended mind thesis to include 'mental institutions' such as educational practices and entire legal systems (Gallagher 2013; Gallagher and Crisafi 2009).

What is important about this approach from the perspective developed in this book is that it offers a clear example of how the already strong sense of the transcendental manifest in 4e work might be developed further. Consider four points:

1. There is nothing in Gallagher and Crisafi's approach that rules out its compatibility with a general commitment to methodological naturalism that is exhibited right across work in the 4e context (see, for instance, Clark 1996). This is because Gallagher and Crisafi are not committed to positing any supernatural or transcendent entities or processes.[27]
2. There is nothing in Gallagher and Crisafi's approach, in principle, that rules out its capacity to act as a framework for multifarious and fine-grained empirical case studies (of particular educational practices, legal precedents or technological systems, for instance).
3. Gallagher and Crisafi are simply committed to extending our sense of the scope of conditions further, to incorporate entities and processes that lie outside more restrictive epistemological purviews.[28]
4. A crucial consequence of their approach, if taken seriously, is to open crossover potentials for work between 4e and other areas of contemporary philosophy.

In an attempt to emphasize the cumulative force of these points, and by way of drawing the claims advanced over the course of this chapter together, let me dwell on the fourth point. In a passage worth citing at length, Gallagher puts a version of this point as follows:

> What I suggest is twofold. First, that the concept of the extended mind ... offers a new understanding of what cognition ... actually is and how it works. As such it offers a new perspective for understanding decision making, judging, problem solving, communicative practices, and so forth, which importantly includes reference to the kind of externalities that critical theory ought to be concerned about – institutional practices and procedures, norms, rules, technologies. ... Such externalities not only shape our cognitive processes and thinking, but also play a dominating role in bureaucratic systems, democratic processes, and in an extensive range of social, legal, and political phenomena. ... The idea of the socially extended mind at the very least offers a new tool for the practice of critical theory. Second, although cognitive science is already studying the kind of cognition that some theorists take to be socially extended ..., the proposal here is that we give this kind of cognitive science a critical twist. (2013: 9)

In the previous part of this chapter, I argued that the concept of the extended mind draws on a strong sense of the transcendental, to, as Gallagher puts it, '[offer] a new understanding of what cognition is and how it works'. I further argued that, when situated within the broader context of 4e work, this sense of the transcendental becomes even stronger and more multifarious. This, I argued, can be seen in debates concerning constitutive ambiguities in the extended mind thesis, as well as in the differing emphases that approaches emerging from the 4e context put on conditions of 'embodiment', 'extension', 'enaction' and 'embeddedness'.

A key consequence of these issues, I claimed, is that work in the 4e context can be viewed as offering a refined taxonomy for complementing approaches to embodiment conditions emerging from continentally inspired philosophy of technology (as considered in the case of Dreyfus in Part 1 of this chapter) and media theory (as considered in the cases of Hayles and Hansen, in Part 2). What should be emphasized in light of Gallagher's comments in the above passage, however, is that this cuts both ways. This is because philosophy of technology and media theory are fields where the types of critical theory issues Gallagher touches on are at stake in pronounced and specific ways, in terms of case studies that differ from what Gallagher (albeit contentiously) identifies as the set of 'typical examples ... rehearsed in the extended mind literature' (2013).[29]

The upshot of this is that just as there is considerable potential for philosophy of technology and media theory to draw on the precision of analytical tools developed in the 4e context, so too is there considerable potential for 4e work to draw on philosophy of technology and media theory, in ways that take up Gallagher's call for 'cognitive science [with] a critical twist'.

To be clear, this is not to claim that the concerns of philosophy of technology, media theory or 4e work are reducible to one another. Instead, it is simply to note that there are considerable potentials for crossover work between these areas.[30] What makes this work possible, in part, is a shared sense of the transcendental that work in each of these areas has, as I hope to have shown in this chapter, to date focused on embodiment conditions to good and multifarious effect. Echoing Gallagher and Crisafi's claim in favour of the socially extended mind thesis, however, a developed and dynamic sense of the transcendental teaches us something else: to aspire to be as open and reflexive as possible when considering *conditions*.

What this implies is that work exhibiting a sense of the transcendental across philosophy of technology, media theory and 4e is compatible with attempts to engage conditions that currently lie outside the scope of some of the more restrictive epistemological purviews operating in each of these areas. Such conditions are at stake in Gallagher and Crisafi's concept of 'mental institutions', and they are also at stake in the concept of exceptional technologies, understood in terms of artefacts and practices that exceed and challenge our received sense of what constitutes a technology. In inquiring after these conditions, we need not lose a concern for embodiment conditions, and we need not attempt to reduce the concerns of philosophy of technology, media theory or 4e work to one another. Instead, we can play on the crossover potentials between these fields. In this part, I have attempted to build the case for how the compatibility just highlighted works in principle. In the next chapter, I will attempt to show how it works in practice, through three case studies of exceptional technologies.

CHAPTER FOUR

Three exceptional technologies

Towards the end of Chapter 1, I claimed that 'exceptional technologies' can be defined as artefacts and practices that appear as marginal or paradoxical exceptions to a received sense of what empirically constitutes a technology in a given context of design, implementation or use, but that can nevertheless act as important focal points for drawing out and challenging conditions implicated in the received sense. Examples of such exceptional technologies include ostensibly trivial, merely imagined, failed and impossible technologies, and the range of conditions they allow us to focus upon can, for instance, be political, aesthetic, economic, logical, epistemological and ontological. In Chapter 2, I attempted to show how an ostensibly trivial technology (the blank page) can be viewed as exceptional in this sense. Chapter 3 then worked towards the claim that an approach focused on exceptional technologies is compatible with a sense of the transcendental emerging from recent approaches to embodiment conditions in philosophy of technology, media theory and work in '4e' philosophy of mind and cognitive science.

But what about merely imagined, failed and impossible technologies? How can these act as focal points for drawing out conditions implicated in actual processes of design, implementation and use? The aim of this chapter is to respond to this issue through case studies of each of these types. In Part 1, I focus on the case of Vannevar Bush's 'memex', a merely imagined technology that has nevertheless had important and well-documented influences on developments in networked digital computing since the second half of the twentieth century. Part 2 focuses on Francis Galton's controversial practice of 'composite photography', a failed technology according to Galton's own deeply problematic standards of success, but one that raises important issues for contemporary work in areas including facial recognition technologies, bioimaging and data visualization. Part 3 then focuses on Arthur Ganson's 'Machine with Concrete', a work of kinetic sculpture with a self-consciously impossible aim.

It should be noted from the outset that each of the case studies considered in this chapter has a long history across various fields: Bush's memex is well-known in media theory (Chun 2008; Gere 2008; Frieling 2004), Galton's composite photography is well-known across fields including cultural studies, the history of photography and the history of science (Watts et al. 2008; Sekula 1986; Hacking 1990), and Ganson's work is well-known in the contemporary art world, with a long-running exhibition of his sculptures at MIT and popular videos of his work on YouTube and TED (MIT 2017; Ganson 2004, 2008). My aim in citing these case studies, then, is by no means to lay claim to them as the proper objects of philosophy of technology, as if this constituted a 'master-discipline', capable of subsuming the others. Instead, it is to show how we might draw on a range of different disciplinary perspectives to inform work in philosophy of technology, and to show how the concept of exceptional technologies may be of cross-disciplinary interest.

It should also be noted that the case studies of this chapter are in no way offered as a definitive set of 'exceptional technologies'. Instead, my aim is simply to provide some more content for a concept that may, I hope, prove to have much greater extension. By this, I mean that there may be many more examples of merely imagined, failed, and impossible technologies that could have been preferred to the examples considered in this chapter, and that there may be many other types of exceptional technologies than the terms 'merely imagined', 'failed' and 'impossible' cover.[1]

Let me also emphasize a methodological point here: the case studies of this chapter can, I hope, either be read as a whole or as stand-alone pieces. When read in the former sense, in terms of the general drift of the argument of this book, what is fundamentally at stake in these case studies is this: the very possibility of such a thing as a case study of an 'exceptional technology' for an expanded and cross-disciplinary conception of work in philosophy of technology. By demonstrating that such a thing is possible, what this chapter seeks to show is that the empirical turn in philosophy of technology has not so much been incorrect, as *insufficiently empirical*. By this, I mean that philosophy of technology should, as per the emphasis of its empirical turn, certainly keep its aspiration to focus on detailed case studies of technologies in contexts of design, implementation and use; however, there is no reason why this turning towards the empirical has to occur at the price of a turning away from 'transcendental' concerns regarding conditions. On the contrary, what the case studies of this chapter seek to show is that it is both possible and desirable to go much further in empirical and transcendental directions at once, in favour of a dynamic and thoroughgoing approach to philosophy of technology.

1 Everything but the network: Vannevar Bush's Memex

In July 1945, an enigmatic essay appeared in the US popular magazine, *The Atlantic Monthly*.[2] It was called 'As We May Think', and it was written by Vannevar Bush, one of the foremost US engineers of the day. The essay considered a broad but timely question: with the Second World War approaching an end, how should the United States make peacetime use of its technological and scientific resources?[3] Bush responded that the aim should be to make humanity's collective memory or 'record of ideas' more accessible (Bush 2017: 2). What really caught the imagination about Bush's essay, however, was the machine he proposed for accessing the record. He called it the 'memex' (see Figure 1), and, in describing it, gave subsequent generations strong reasons to think he had predicted developments in personal computing in the latter part of the twentieth century in virtually every key respect.

The aim for this part is to read Bush's memex as an exceptional technology. In doing so, I will focus on this anomaly: while there are indeed many superficial similarities between the memex and contemporary computing devices, Bush's essay demonstrates no conception whatsoever of networking between devices, arguably the most important condition shaping our contemporary experience of information and communication technologies (ICTs).[4] What is exceptional about the memex, then, is this: it is a merely imagined technology, the well-documented legacy of which has led us to overlook crucial differences between it and the networked technologies it helped to inspire.

The original *Atlantic* version of 'As We May Think' features eight sections. Sections 1–4 outline a question (What is science to do after the war?), offer an answer (Science should improve access to mankind's 'record of ideas'), and describe some of what Bush calls the 'new and powerful instrumentalities' that might aid the task, including dry photography, microfilm, stereoscopic film, fax transmission, television, typewriters, and the 'vocoder' (Bush 2017: 2–5). Despite their broad scope, the tone of these sections is measured, striking a balance between the register of the expert and examples that are intended to be accessible enough for *The Atlantic's* lay readership.

It is in section 5 that Bush's essay really gains momentum. First, he identifies a problem: while the technologies described up to this point mean that 'we can enormously extend the record', the record seems too vast to be consulted, 'even in its present bulk' (Bush 2017: 8). Bush calls this the problem of 'selection', and it is to find a solution to it that he devotes the remaining sections of his essay to speculations on the memex.

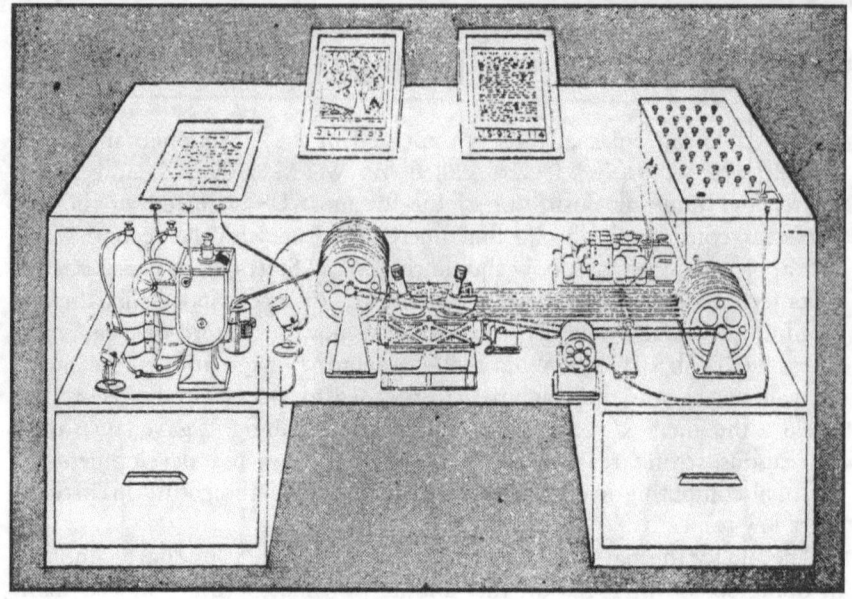

FIGURE 1 *The 'memex', based on Bush's description in 'As We May Think'.*

Here is how section 6 introduces the device:

Consider a future device for individual use, which is a sort of mechanized private file and library. It needs a name, and, to coin one at random, 'memex' will do. A memex is a device in which an individual stores all his books, records and communications [on microfilm], and which is mechanized so that it may be consulted with exceeding speed and flexibility. It is an enlarged intimate supplement to his memory. (Bush 2017: 10)

Bush goes on to elaborate a host of further features of the memex: 'it consists of a desk'; 'while it can presumably be operated from a distance, it is primarily the piece of furniture at which [the individual] works'; there are 'translucent screens ... on which material can be projected for convenient reading'; 'there is a keyboard'; 'only a small part of the ... memex is devoted to storage, the rest to mechanism'; storage capacity is, however, huge, such that the user 'can be profligate and enter material freely'; 'contents are purchased on microfilm ready for insertion'; 'on ... top of the memex is a transparent platen', through which hard copies of texts and images can be 'photographed onto the next blank space'; there are 'levers' and 'mnemonic codes' that allow for quick consultation of contents; there is a 'special button [that] transfers [the user] immediately to the first page of the index'; the efficiency of the memex is such that 'any given book of [the user's] library can ... be called up and consulted with far greater facility than if it were taken from a shelf' (Bush 2017: 10).

These details have proven deeply suggestive for successive generations of readers. With only a short stretch of the imagination, a contemporary reader may, for instance, be inclined to read the following and more into Bush's words: desktop and laptop computers; monitors; the entire history of data storage (from the hard disk to the floppy disk, from the CD to the USB, from the DVD to cloud computing); scanners, photocopiers and printers; hotkeys and shortcuts; homepages; search engines and e-reader software.

Does this mean we should view the memex as a 'proto-ICT'? This is certainly the impression we would get if we stopped reading at section 6 of Bush's essay. There is, however, a key passage towards the end of section 7 that reminds us we may be overlooking something important: nothing in Bush's description parallels networking between devices.

Bush intends section 7 of his essay to give practical examples of the memex in use. The 'essential feature', he tells us, is to allow individuals to 'build associative [encyclopaedic] trails' among the diverse materials inserted into the machine. Towards the end of section 7, Bush gives an example of how such a trail might be exchanged between memex owners. First, he imagines an individual who has drawn on various articles to build an associative trail on the subject of Turkish archery. Next, he imagines this situation:

> Several years later, [the individual's] talk with a friend turns to the queer ways in which a people resist innovations, even of vital interest. He has an example, in the fact that ... outraged Europeans ... failed to adopt the Turkish bow. In fact he has a trail on it. ... He sets a reproducer in action, photographs the whole trail out, and passes it to his friend for insertion in his own memex. (Bush 2017: 11)

Our focus here should fall on six simple words: 'and passes it to his friend'. What these words indicate is that 'As We May Think' envisages no such thing as networking between memexes. Instead, Bush imagines memexes as discrete analogue/mechanical devices, and presupposes that they will each depend on the forms of face-to-face interaction conventional in a given human society for information to be exchanged between them.

In the ten to twenty-five years after 'As We May Think' was first published, profound empirical and discursive changes took place in computing and information science. First, there was a shift away from the analogue technologies envisaged by Bush, towards digital. Second, there was a shift from mechanical to fully electronic systems. Third, there was a shift towards a more coherent vision of 'computer science' (for which Bush's own work in applied computing acted as an important preliminary). Fourth, there was a shift towards the register of cybernetics, which avoided positing 'essential differences' between persons and machines, and instead viewed both alike as 'feedback' nodes in communication networks (Wiener 1968: 18). Fifth, the 'problem of selection' which had prompted Bush's speculations became subsumed within emergent problems of 'communication' and 'networking' (see Triclot 2008).

Throughout all these changes, however, the memex remained what has been termed a powerful 'image of potentiality' (Smith 1991: 262).[5] Nowhere is this more apparent than in Bush's own writings, which retained a focus on the memex for the remainder of his life. In an unpublished piece, 'Memex II', written in 1958, Bush projected two further iterations: a 'Memex II' that would incorporate voice recognition, as well as developments in digital computing and machine learning, and a 'Memex III' that would, Bush speculated, bypass the interface of language entirely, in favour of direct links with the central nervous system (Bush 1991a: 165–84). In 'Memex Revisited', published in 1967, Bush toned down these speculations; he did so, however, while accentuating what he saw as the increased plausibility of the second iteration of his device:

> When ['As We May Think'] was written, the personal machine, the memex, appeared to be far in the future. It still appears to be in the future, but not so far. Great progress, as we have seen, has been made in the last twenty years on all the elements necessary. Storage has been reduced in size, access has become more rapid. Transistors, video tape, television, high-speed electric circuits, have revolutionized the conditions under which we approach the problem. Except for one factor of better access to large memories, all we need to do is to put the proper elements together – at reasonable expense – and we will have a memex. (Bush 1991b: 215)

It was by no means only for Bush, however, that the memex figured as an 'image of potentiality'. On the contrary, it inspired successive generations of computer researchers across an extensive and multiform body of work.

Surveying this literature with retrospect, it is possible to detect a number of trends. From the late 1950s onwards, 'As We May Think' began to be cited as a 'starting point of modern information science' (Smith 1991: 264–5). In the 1960s and 1970s, it was linked to themes of 'Man-Computer Symbiosis' and human-computer 'augmentation', thanks to chains of influence between Bush's essay and the work of eminent figures in US computing, including J. C. R. Licklider, Claude Shannon, Ted Nelson and Douglas Engelbart.[6] Between 1981 and 1990, citations of 'As We May Think' increased twofold, due largely to the fact that Bush's 'trails of association' were now being cited as a significant precursor to Nelson's concept of hypertext (Smith 1991: 265; Houston and Harmon 2007: 66). References to the memex then unsurprisingly grow exponentially from the 1990s, in fields as diverse as literary theory, library science and marketing, thanks to the emergence of the World Wide Web and, from the late 1990s, to rapid developments in search (Bolter 2000; Houston and Harmon 2007: 77–82).[7]

Today, the memex's direct influence is perhaps most visible in the field of knowledge management (Houston and Harmon 2007: 55), and in design problems surrounding 'Personal Knowledge Bases' (PKBs) (Davies 2011). In the contemporary context, a PKB can be defined as 'an electronic tool through which an individual can express, capture, and later retrieve the personal knowledge he or she has acquired' (Davies 2011: 81). In other words, a PKB acts as a personal epistemic filter for an individual immersed in a networked culture.

With this emphasis on the 'personal' and the 'individual' in mind, let us now return to a consideration of the specifics of the memex, as outlined in 'As We May Think'. Why, as section 6 of 'As We May Think' suggests, was Bush able to foresee so many of the details of the human experience of computing since the latter part of the twentieth century, while, as section 7 clearly indicates, he failed to foresee networking between devices, perhaps the most important conditioning factor?

A simple answer would be that Bush had no conception of the empirical plausibility of such networking. That, however, would seriously underestimate knowledge demonstrated over the course of his essay, and in subsequent writings. Specifically, it would underestimate his description of the 'telautograph', a device for transmitting signatures between telegraph stations (Bush 2017: 6), as well as the automatic telephone exchange, a form of network described twice in 'As We May Think', but not connected to the idea of the memex (Bush 2017: 3, 8). Moreover, it would ignore albeit vague remarks that Bush makes in 'Memex II', on the possibility of networking by telephony to 'massive memexes' in central libraries (Bush 1991a: 173–4).

A more plausible answer might concern the conceptual conditions under which 'As We May Think' was written. What if Bush failed to foresee the subsequent importance of networking, at least in part, because of big philosophical presuppositions concerning the nature of thought, subjectivity and memory that are demonstrated over the course of his essay?

There is a key passage in 'As We May Think', immediately before the memex is introduced, that gives important clues on conceptual conditions informing the device. Having criticized the inefficiency of existing alphanumerical index systems for accessing the 'record', Bush states:

> The human mind does not work that way. It operates by association. With one item in its grasp, it snaps instantly to the next that is suggested by the association of thoughts. ... [The] speed of action, the intricacy of trails, the detail of mental pictures, is awe-inspiring beyond all else in nature. ... Man cannot hope fully to duplicate this mental process artificially, but he certainly ought to be able to learn from it. In minor ways he may even improve [it], for [artificial] records have relative permanency. ... It should be possible to beat the mind decisively in ... the permanence and clarity of the items resurrected from storage. (Bush 2017: 9–10)

What key presuppositions are revealed by this passage, which was central enough to Bush's thinking to reappear almost verbatim over twenty years later in 'Memex Revisited' (Bush 1991b: 198)? Of thought, Bush presupposes a basic 'associationist' model of the psyche.[8] Of subjectivity, he presupposes that 'man' will remain firmly in control of any machine he might create, since he will always be capable of 'thinking better' than it (Bush 2017: 12). Of memory, he presupposes that it can be externalized and delegated to a machine, but, again, crucially, that man will remain the agency in control of this process.

The key point about these presuppositions is that, taken collectively, they serve to inhibit recognition of a host of issues likely to be generated by networking between devices. Technologically, Bush's emphasis on centralized individual control inhibits recognition of distributed networks, as well as the forms of ubiquitous, mobile and miniaturized ICT use for which networking has acted as an indispensable condition of possibility (Weiser 1991; Galloway 2004; Chun 2011). Economically and sociologically, Bush's emphasis on the memex as a bespoke and non-networked device inhibits recognition of the potential for networked ICT use on a massive scale, as made possible by developments in consumer computing and the microprocessor revolution of the 1970s and 1980s (see Watson 2012: 125–59). Epistemologically, Bush's account demonstrates no conception of issues posed by search engines since the latter part of the twentieth century, such as the increased importance of automation and algorithms in the management and generation of knowledge, issues surrounding open access, or the potential for the commercialization and gamification of page ranking.[9]

The point here is not to condemn Bush for failing to foretell the future; it is to draw attention to the limits of any contemporary inclination to think that he did. This is because the presuppositions underpinning the memex push in a direction contrary to a horizon of issues opened up by networking. Instead of looking outward to this horizon, Bush's presuppositions push 'As We May Think' inward, towards a form of methodological individualism or solipsism that takes the putatively associationist constitution of the individual's brain to be the appropriate criterion for the management of knowledge.[10] In turn, this sets the scene for a very specific and limited example: a machine that makes and stores associations (i.e. 'thinks') in a clearly defined and self-contained way, and that is envisaged to remain firmly under the control of a single owner.

As imagined by Bush, the memex is precisely this device, and what is further revealing in this respect is the profile Bush assigns to the average user. Beyond researchers like himself involved in the 'hard sciences', he cites lawyers, attorneys, physicians and historians as prospective memex owners, as well as what he suggestively calls a 'new profession of trail blazers' (Bush 2017: 12). In other words, Bush presupposes 'memexes' to be specialist equipment in the hands of a specialized elite, the members of which will each be expert enough to have complete control over the contents of their devices.[11]

What renders the memex an 'exceptional technology', given this picture, I suggest, is the enduring legacy it has enjoyed *in spite* of these issues surrounding its conceptual conditions. Today, there is a view that the memex foretold the future of computing. What this ignores, however, is not merely the fact that the memex was imagined as a mechanical, analogue and non-networked device, but that it was imagined under conceptual conditions that actively inhibited recognition of the kinds of emergent issues posed by networking.

What the case of the memex demonstrates, at the very least, then, is that issues of influence in actual processes of technological design, implementation and use can be much more complex and convoluted than any desire for a clear-cut conception of the 'empirical' as a historically stable category would have us believe. Instead, the case highlights the ongoing need to be attentive in a transcendental sense: to shifts in the conditions under which the limits of the empirical are imagined and reimagined.

To draw this last point out, reconsider Bush's example of a memex user building a trail on Turkish archery. Suppose this was a page on Wikipedia or a post on social media (see Wikipedia contributors 2017). How would this alter the process of knowledge exchange? First, since these sites are subject to varying norms of 'user generated content', the thought process involved in editing the material would aspire to varying degrees of 'publicity'. Second, personal relations with the trail builder would no longer be necessary to access the material. Third, a more thorough externalization of memory would be effected, since responsibility for archiving the page would be delegated away from the user.

An obvious objection here is that Wikipedia and social media are hackneyed examples. This, however, is precisely to the point: such platforms have become trivial and everyday for users of contemporary 'PKBs', by virtue of the ubiquity of networks. A further objection that might be raised is that 'PKBs' that are not 'networked' in an obvious or explicit way are very much still around (paper diaries and journals, for instance). The claim, however, is not at all that these artefacts have become impossible. Instead, it is that their conditions of possibility have significantly changed. Put simply, the kinds of issues to do with networking that were unrecognized possibilities for Bush have today become everyday realities for billions of human beings worldwide, and this has changed the conditions under which all manner of technological, economic, sociological and epistemic practices are possible, whether or not they are materially 'networked' in an obvious and explicit way (see Floridi 2014: 8–9; Castells 2010: xxxi).[12]

When he first introduced the memex, Bush referred to it as '… a future device for individual use, which is a sort of mechanized private file and library' (Bush 2017: 10). The key words here are 'individual' and 'private'. This is because the memex rests on presuppositions that take the nature of human thought, subjectivity and memory to ultimately be of such a character. Whenever the memex is viewed as part of a narrative leading inexorably to our contemporary situation (as it often implicitly is), it may be that we are hearkening nostalgically after these presuppositions, and that we are ignoring emergent issues posed by the forms of 'collective' and 'public' thought, subjectivity and memory that networked cultures make possible. However, another move is open to us: when we focus on what makes a case like that of the memex an *exception* to simplified narratives, it may help us loosen engrained desires to simplify technology's role in shaping history, in favour of a less diminished sense of the contingent possibilities and challenges that technologies pose for the future.

2 'Pictorial statistics': Francis Galton's composite photography

Let us now move to consider a different case study: Francis Galton's 'composite photography'. This practice developed as a supplement to Galton's work in statistics and, most notoriously, to his work in eugenics.[13] Introduced in 1877, composite photography remained an important part of Galton's work until his death in 1911. For him, the practice confirmed the theory of racial and class types underpinning his eugenics. Today, beyond its importance as a case study for the history of art and photography, Galton's practice forms a controversial part of the history of disciplines including anthropology, criminology and biometry (Maxwell 2008; Sera-Shriar 2015; Wade 2016).

The aim of this part is to read Galton's composite photography as an exceptional technological practice. I will focus on the following anomaly: starting out from what he took to be rigorously empirical premises, Galton's practice produced speculative images of 'types' that were unverifiable and unfalsifiable. What is exceptional about Galton's practice, then, is this: it is, by its own initial empiricist standards of evidence, a practice that *must* fail to meet its aim. In turn, this failure raises important issues for areas of contemporary research including facial recognition technologies, bioimaging and data visualization.[14]

Here is how Galton described his practice:

> My method of composite portraiture ... consists in throwing the images of different pictures successively upon the same screen, giving to each a proportionate fraction of the total length of exposure required to produce an ordinary photograph; the result being that what is common to all the pictures has been adequately exposed and is retained in the resulting photograph, and what is individual to each of them has been too under exposed to leave any image at all, and consequently disappears. (1900: 135)

Galton's method typically involved synthesizing between five and ten component photographic portraits of the human face (see Figure 2). Galton did, however, produce images with as few as two components and as many as a hundred, and also experimented with materials including Eadweard Muybridge's famous 'Horse in Motion' series (Ellenbogen 2012: 124– 8; Sekula 1986: 45; Galton 1882). From 1877 to 1888, when working

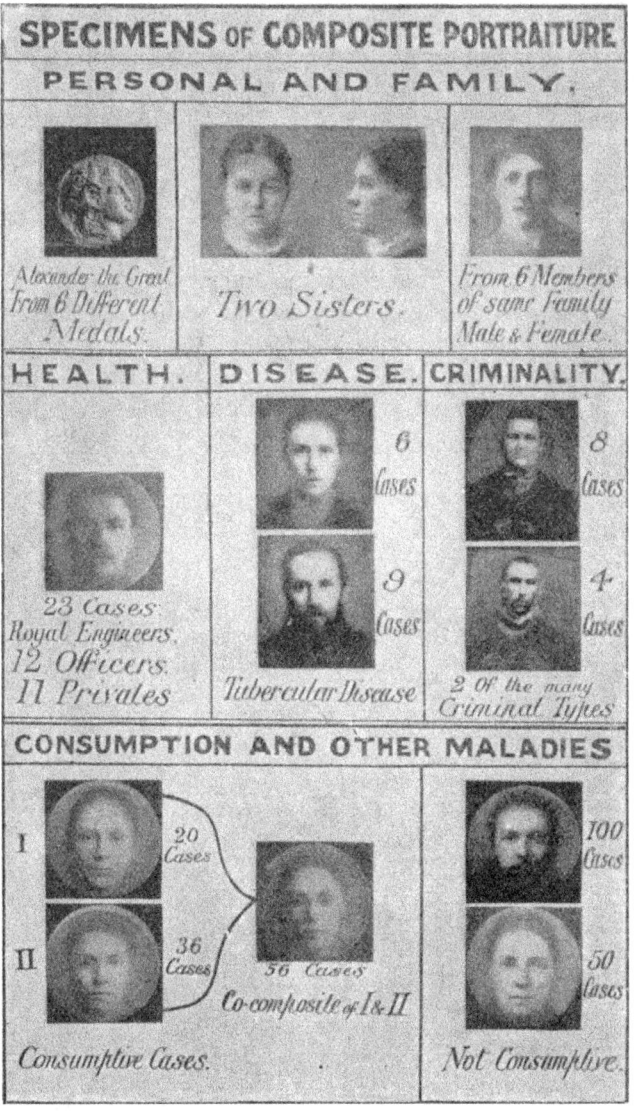

FIGURE 2 *Examples of composite photographs from Galton's* Inquiries into Human Faculty *(1883).*

most intensively on his practice, Galton developed his technology from a basic wooden rig requiring portraits of relatively fixed scale to a more sophisticated backlit apparatus with a zoom (Figures 3 and 4). Galton's basic method remained consistent throughout this time, however. First, a pack of photographic portraits was hung in front of a high-end camera. Next, the pack was framed by a brass grid for consistency of framing (Figure 5). Successive portraits were then superimposed upon one another by removing the camera's cap for a consistent exposure time, and shifting through the pack. In an 1878 lecture to the British Anthropological Institute, Galton mooted an exposure time of 10 seconds per image in the pack (1879a: 133–4).

What remained even more consistent about Galton's practice over this time was its aim. As already indicated, he wanted to produce composites that eliminated differences to reveal identical 'types' (Figure 2). In his 1878 lecture, Galton put it like this:

> [A composite portrait] represents no man in particular, but portrays an imaginary figure possessing the average features of any given group of men. These ideal faces have a surprising air of reality. Nobody who glanced at one of them for the first time, would doubt its being the likeness of a living person, yet, as I have said, it is no such thing; it is the portrait of a type and not of an individual. (1879a: 133)

FIGURE 3 *Galton's basic wooden rig, c. 1878.*

FIGURE 4 *Galton's more specialized apparatus, c. 1881.*

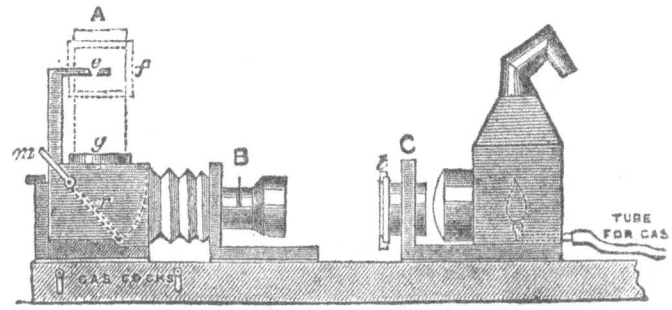

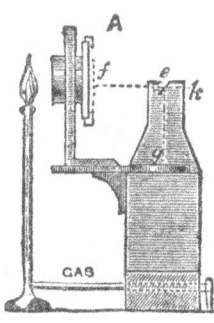

FIGURE 5 *Detail of Galton's brass framing grid, c. 1878.*

Galton's big claim in this passage (that composites are 'portrait[s] of a type and not of an individual') forms the key interest of his practice for him. It is also, however, what gives rise to the most controversial philosophical issues surrounding his practice. First, there is an issue of circularity: to produce the composite of a type, it was necessary for Galton to first presuppose a concept of that type as a way of identifying suitable component images. Second, there is a perennial philosophical issue concerning realism and nominalism. This concerns whether Galton took types to exist in nature, or to be constructed and conventional, and, as we will see, it was an issue on which his practice was problematically ambivalent. Third, big issues concerning ideology and power are at stake in terms of how types get selected for a process like this. In Galton's case, types selected included 'the Ideal Family Likeness', 'Brothers', 'the Criminal', 'the Consumptive' and 'the Jewish Type' (see Figure 2), and this roster, especially when combined with knowledge of Galton's work in eugenics, has left him open to well-placed criticisms that his practice simply served crude ideological interests (whether patriarchal, anti-Semitic, or biased towards the interests of a narrow 'professional middle class' (Sekula 1986: 40; Green 1985: 14)).

The issue I want to focus on in this part relates to all of these other issues, which are well-covered in the existing literature on Galton.[15] It concerns Galton's relationship with empiricism. Up to a point, Galton's practice was based on an apparently sound empirical understanding of what were, at his time, cutting-edge photographic technologies and experimental methods. Beyond this point, however, it tended towards extraordinary excesses. What's more, it did so on the strength of Galton's commitment to a form of empiricism. Identifying where this point lies is therefore crucial for considering what marks Galton's practice out as an 'exceptional technology' in the sense developed in this book.

There are at least three aspects of Galton's practice that might have lent it the air of an empirically sound scientific practice to his contemporaries. First, Galton's results were, he claimed, replicable.[16] Second, Galton favoured the most expensive and up-to-date photographic equipment throughout the development of his work (Sera-Shriar 2015).[17] Third, Galton's component images bore what his contemporary Charles Sanders Peirce famously described as an 'indexical' relation to their objects: as photographs, they attested to the real physical existence of the subjects who sat for them.[18]

What undermined these credentials, paradoxically, was the strength of Galton's commitment to a late form of British empiricism. Drawing on the work of his friend Herbert Spencer, as well as that of Thomas Huxley and David Hume, Galton went so far as to take the photographic process of his day for a model of the process whereby general ideas of 'type' are formed in the human mind (Ellenbogen 2012: 111; Galton 1879b: 6). This entailed a whole series of problematic equivalences. First, Galton took his component photographs to be equivalent to visual sensory impressions. Second, he took photographic exposure time to be equivalent to the process of vision's habitual exposure to images of a certain type (down, indeed, to an exact equivalence in exposure time (Ellenbogen 2012: 20)). Third, Galton took 'types' themselves to be the equivalents of composite photographs (Ellenbogen 2012: 12).

This is an extraordinarily tendentious account of idea formation, resting on a strained form of analogical reasoning. Even if it is granted, however, a further problem of circularity faces Galton: all composite photography can tell us about on this account is the presupposed process of idea formation itself. What it cannot settle is the issue of whether ideas of type have real or merely 'nominal' existence. This is because, on Galton's chosen empiricist model, the same process of idea formation holds for all ideas, whether they refer to really existing entities or constructed fictions. For Hume, for instance, fictions such as the idea of a 'Golden Mountain' are formed in the mind as complex ideas through a process of blending simpler component ideas (2007: 18–19), and what is problematic about this for Galton, in turn, is this: even if we grant strict parallels between his practice and a Humean account of idea formation, this still leaves him with no criterion by which to distinguish the products of his practice from such fictions.

This points us towards the most tendentious aspects of Galton's method: he appears to have taken his component images, by virtue of their indexicality, to have in fact provided a criterion for distinguishing fact from fiction. What is mistaken about this, however, is that such a criterion could not carry across to his composites: instead of being 'Indexes', these were what Peirce called 'Symbols'.[19]

To draw out what is at stake in this distinction, consider the following from one of the best recent accounts of Galton's work, from Josh Ellenbogen:

> Galton tried to make photographs of ideas, type concepts that embraced all the individuals of a class. Such images clearly have a difficult position relative to what we see. While the human eye can behold individual horses and individual criminals, the eye can never see, in its experience of the natural world, species types such as 'the horse' or 'the criminal', two of the ideas Galton frequently sought to present. ... What standards govern a picture that aims to depict an idea? (2012: 7–8)

The deep problem facing Galton is that he could provide no clear answer to this question. On the one hand, his composites could not bear an indexical relation to 'types', because, as Ellenbogen highlights, there are no such visible objects that we can claim to directly experience in the world in any unproblematic sense. On the other hand, there was an obvious class of visible objects to which Galton's composites could be compared: artworks designed to display standards of type.[20] The problem facing Galton, however, is that any such comparison would only be governed by conventional standards of symbolism, because no comparable artwork could itself claim an indexical relation to the visible object 'type'.

Although Galton's composites may conceivably have been replicable given similar component photographs, apparatuses and practical conditions, then, the lack of an appropriate visible object 'type' to which they could be indexed meant that his claim that they were 'portrait[s] of a type and not of an individual' was neither verifiable nor falsifiable. This, moreover, was a logical constraint on Galton's practice, and not a contingent problem to be resolved by methodological fine-tuning: by the standards of the empiricist epistemology on which Galton modelled his account of photography, general ideas of type, while formed by habit, are not themselves visible objects; instead, they are conceptual conditions under which objects (including visible ones) get individuated and identified in experience. No matter how technologically refined Galton's practice became, then, his aspiration to produce 'portrait[s] of a type and not of an individual' was destined to fail by his own initial empiricist standards of evidence.

Why and how, given this picture, should Galton's composite photography be considered an 'exceptional technology' that has anything instructive to say about wider processes of technological design, implementation and use today? On the one hand, it would be easy to recognize the case as an idiosyncrasy or historical curiosity, and to leave things there. What this would overlook, however, are instructive resonances that exist between the case and work in many areas of contemporary research, including into facial recognition, bioimaging and data visualization technologies.

In moving to draw out these connections, it is particularly important to be wary of committing a version of the 'genetic fallacy'.[21] As we have seen, Galton's practice is surrounded by all kinds of problematic ideological issues, and can rightly be described as a failed practice because of its inability to live up to the standards of scientific proof to which he submitted it. It would be a mistake to conclude from this, however, that any work building upon comparable methods of photographic superimposition and synthesis must logically be affected by these issues.

Instead, I want to use the remainder of this part to develop a point that runs counter to an impulse towards the genetic fallacy. What is most instructive about Galton's composite photography, I suggest, is precisely that its methods turned out *not* to be reducible to the values he invested in it, or to the applications to which he put it: Logically speaking, it is not the case that work building upon comparable methods of composite photography *must* be subject to the kinds of bad presuppositions operating in Galton's case; what Galton's case offers precisely because of this, however, is the opportunity for a lateral and instructive shift that allows us to foreground just some of the many ways that comparable practices *can* fall victim to bad presuppositions.

To draw these issues into sharper relief, consider the following remarks on Galton in a recent essay by Suzanne Bailey:

> Galton's composites ... have the effect of creating a tear in the cultural imaginary, by creating new possibilities for visual representation in addition to whatever ideological function or scientific purpose they may originally have had. In Galton, there is a residue of a ... naïve realism in his hope that producing images of the human face in composite layers would reveal scientific laws or principles. (2012: 197)

Bailey's language of a 'tear' may seem hyperbolic here, but it is appropriate.[22] Her point is that Galton's experiments with composite photography were introducing new and 'surreal' possibilities into facial recognition at a historical point when many already felt intimidated by basic photographic portraits (2012: 197). Out of this tear, Bailey proceeds to note methodological and conceptual resonances between Galton's practice and new possibilities for visual representation that have since been realized, including 'dissolve transition' (2012: 191), visual search engines (2012: 195), digitized images (2012: 197), Photoshopped artworks (2015: 203), forensic applications designed to show aging in a face over time (2012: 209), and politically motivated work by artists seeking to directly counteract Galton's emphasis on 'type' (2012: 211).[23]

If we push this line of inquiry further, three broad areas of contemporary research emerge as resonating particularly strongly with issues raised by Galton's practice: work in facial recognition technologies, bioimaging and data visualization.

At first glance, facial recognition technologies seem to deal with a problem that runs in the opposite direction to that of Galton. Whereas he combined individual portraits to generate images of 'type', computer scientists developing algorithms for automated facial recognition systems today are usually concerned with how to identify individuals (Gates 2011: 19–24). Again, bioimaging technologies such as ultrasound, fMRI or CT scanners seem to deal with distinct issues. Whereas Galton mistakenly sought to produce images of ideal objects, bioimaging technologies deal with material objects on scales ranging from the nano- and the molecular to macroscopic entities such as the unborn child or the flow of blood in the brain, across different regions of the electromagnetic spectrum (Vadivambal and Jayas 2016). Lastly, the implications and applications of modern data visualization techniques like scattershots, parallel coordinate plots or clustered heatmaps seem remote from Galton's practice. Galton positioned his composites as a means of directly perceiving 'type', with the implication that details could be overlooked (Pearl 2010: 203). In contrast, data visualization is used in fields such as big data analysis and bioinformatics to address epistemological problems generated by the complexity of large datasets: rather than eliminating detail, data visualization in these fields seeks to preserve it in accessible forms that can subsequently be put to a variety of analytical uses (Friendly 2008; Galloway 2012: 78–100).

What renders Galton's composite photography an instructive case in spite of these clear differences, I suggest, is its capacity to offer a lateral move for focusing on issues affecting these fields in a way that is extreme, yet accessible. As merely the short survey just given indicates, the complexity of issues generated by these fields can be formidable. What is instructive about a case like that of Galton's composite photography in the face of this, then, is that it offers a way of cutting through this complexity to focus on core philosophical issues affecting these fields.

Consider facial recognition technologies. If we suspend a concern for whether these tend towards the identification of 'types' or individuals, they must, like Galton's practice, presuppose preselected databases of images. This means that computer scientists working in this field today, like Galton, encounter issues of circularity concerning the criteria by which their databases get populated (Gates 2011: 19–24). What's more, contemporary computer scientists encounter these issues on a massively more complex and more technologically mediated scale (hence their recourse to algorithms). As noted above, there is no logical connection between this complexity and a tendency to fall into bad presuppositions on the uses to which facial recognition technologies might be put (for socially repressive or racist purposes, for instance). What Galton's case offers in the absence of such necessity, however, is something much more valuable: a way of focusing on the always present contingency that such presuppositions might arise to affect the inputs and outputs of work in this field.

Now consider bioimaging. The images generated by technologies in this field, like those of Galton's practice, presuppose interpretation, and this raises acute issues when considering contemporary tendencies towards large datasets and automation. In studies in bioinformatics, for instance, phenomena of 'automation bias' have been noted, where clinicians manifest tendencies to place too much faith in interpretations generated by automated clinical decision support systems, leading to errors of both commission and omission in patient diagnosis (Goddard et al. 2012; Mukherjee 2017). What Galton's case offers in the face of this is a lateral shift away from factors conditioning this form of bias (such as the complexity of large datasets and presuppositions concerning the perceived epistemic superiority of automated systems), in favour of a focal yet extreme example of how even small datasets can generate bad interpretations.

To now conclude this part, let me focus on the case of data visualization in a little more detail. Recall a dilemma we left Galton facing above: he wanted his composites to be governed by indexical standards, but could not have them, and he could have symbolic standards, but did not want this to relegate his work to the status of a merely conventional 'art practice'. What should be noted here is the way that Galton sought out of this: by interpreting his composites in terms of a different kind of symbol. Rather than symbolizing merely conventional artistic standards, Galton took his composites to depict deeper statistical patterns governing the natural evolution of 'types'.

In an 1879 lecture to the Royal Institute, Galton stated:

> The process of composite portraiture is one of pictorial statistics. ... Composite pictures are ... much more than averages; they are rather the equivalents of those large statistical tables whose totals, divided by the number of cases, and entered in the bottom line, are the averages. They are real generalisations, because they include the whole of the material under consideration. The blur of their outlines, which is never great in truly generic composites, except in unimportant details, measures the tendency of individuals to deviate from the central type. (1879b: 5–6)

The kind of tables Galton had in mind here were Gaussian error tables or 'bell-shaped curves' (see Figure 6). These had been used by the incipient social sciences of Galton's time to measure the tendency of values such as height and weight to evolve towards a mean in given human populations, and to drop away in symmetrical binomial gradations towards the extremes (Sekula 1986; Hacking 1990). Just as in the case of Galton's equivalence between idea formation and photography, the equivalence he was making between his composites and such tables implied a whole series of further

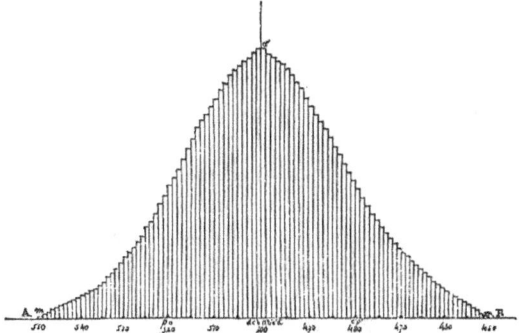

FIGURE 6 *Binomial distribution.*

problematic ones: first, he took his indexical component images to be equivalent to statistical data; second, he took the repeatable features emerging from a 'truly generic' composite to be equivalent to the 'central type' (the mean value) emerging in an error curve; third, he took the idiosyncratic non-repeating features, emerging as a composite's 'blur', to be equivalent to the deviant values represented by the curve's extremes.

In a landmark essay on Galton, Allan Sekula draws attention to precisely what is so problematic about these equivalences:

> Galton believed that he had invented a prodigious epistemological tool. … His interest in composite imagery should not be regarded as a transparent ideological stunt, but as an overdetermined instance of biopositivism. (1986: 46)

'Overdetermination', in the sense intended by Sekula here, occurs when more causes are present in an explanatory account than are necessary to determine a given effect (Swain 1979). This in mind, we can observe that Galton's practice was overdetermined in a precise way: Galton, it seems, was intentionally seeking to overdetermine his practice by providing multiple accounts of the causes he took to justify it. As we saw above, he first took his practice to be justified as a literal index of idea formation. Second, he took it to be justified in connection with the symbolic standards of artistic depictions of 'type'. As we are now seeing, Galton also took his practice to be justified in a third way: as a means of visualizing the evolution of statistical patterns of type.

The third of these equivalences is the most 'biopositivistic' in Sekula's sense. This is because it implies an account of vision as a process that is quantifiable and mathematizable, all the way down. Consider, for instance, Galton's claim that 'composite pictures are … much more than averages'. What is implied by this 'much more' is Galton's sense that composites correspond to rule-governed mathematical patterns of type formation. Now consider Galton's claim that composites are 'real generalisations'. The sense of 'real' involved here is material, in contrast to the abstract and symbolic way Galton presupposed error curves to embody truths for statisticians like himself. Reading these two claims together, we can see that Galton took his composites not merely to demonstrate the presupposed truths of error curves, but to perfect their expression (Ellenbogen 2012: 117). On the one hand, he took his composites to point towards principles of a mathematically rule-governed real, underpinning immediate experience (Ellenbogen 2012: 165). On the other, he took them to make this reality palpable in a more natural and immediate way than the usual mathematical symbolism.[24]

The extraordinarily confused claim Galton's third equivalence makes, then, is that his composites offer a better mode of data visualization than the Gaussian error curve. As with his claim for an equivalence between photography and idea formation, this involves issues of circularity and bad analogical reasoning. What it also offers, however, is an extreme example of how data visualization techniques can be co-opted for rhetorical purposes. As Sekula puts it:

> In effect, Galton believed that he had translated the Gaussian error curve into pictorial form. The symmetrical bell curve now wore a human face. This was an extraordinary hypostatization. ... Galton, in seeking the apotheosis of the optical, attempted to elevate the indexical photographic composite to the level of the *symbolic*, thus expressing a *general law* through the accretion of contingent instances. In doing so, Galton produced an unwitting caricature of inductive reasoning. The composites signified, not by embodying the law of error, but by being rhetorically annexed to that law. (Sekula 1986: 48–55, Original emphasis)

As noted above, data visualization techniques are today used to address epistemological problems generated by complexity, while preserving the capacity to zoom in on details. Furthermore, work in this field thrives on semiotic relations of difference rather than similarity (for instance: a clustered heat map in bioinformatics need not resemble the particular disease it charts in a population, and it would be absurd to demand that it should). In contrast, Galton's third equivalence involves eliminating detail and is premised on a reductive and faulty form of analogical reasoning.

What I want to emphasize in concluding this part is this: it is precisely these faults that have the capacity to render a case like that of Galton's composite photographs instructive for the design, implementation and use of data visualization systems today. By differing so markedly from best practice, the case offers a focused example of how tendencies towards simplification and crude forms of pattern recognition can always arise to affect the reception of even the best designed data visualization techniques.

What the case of Galton's composite photography highlights, in this sense, is the inadequacy of any presumption that data visualization straightforwardly resolves epistemological problems of complexity. Instead, even the best designed and most intuitive data visualization systems defer these issues, and, as such, have the capacity to ramify them considerably. What is therefore required in tandem with the development of such systems, as Galton's case starkly highlights, are more refined forms of visual literacy, to identify and work against potential rhetorical abuses and simplifications.

As we have seen in this part, Galton's practice of composite photography must fail by its own deeply problematic criteria of success. Despite this, however, the case is not an idiosyncrasy or an isolated historical curiosity. Instead, it is an exceptional technology that has the capacity to help us to gain important perspective on developments affecting the design, implementation and use of technologies in contemporary fields including work on facial recognition technologies, bioimaging and data visualization. To qualify Galton's practice in this way is not to commit a naïve version of the genetic fallacy. Instead, it is to claim that the case offers something much more instructive and valuable in the absence of any such 'genetic' necessity: a focal case that is extreme yet accessible, and that is helpful for drawing out some of the problematic contingencies that can affect work in the fields just discussed. Given the contemporary tendencies of these fields towards complexity and specialization at increasing rates, moreover, a case study like that of Galton's composite photography is rendered all the more potentially useful. This is because it has the capacity to offer a focus for core philosophical issues, and to contribute to new forms of critically empowered visual literacy thereby (see Galloway 2012: 78–100; Hansen 2015).

3 'Machine with concrete': Arthur Ganson's gestural engineering

Arthur Ganson's 'Machine with Concrete' belongs to a class of entities to which this chapter might have been expected to have devoted more attention up to now: visual artworks. While Bush's memex and Galton's composite photography have had wide-ranging implications for practices across contemporary art and media, neither could be described as primarily 'artistic'. In contrast, 'Machine with Concrete' is a work that builds on the tradition of kinetic sculpture initiated by Jean Tinguely and others in the 1960s (Harris and Lyon 2013: 32). As I will seek to show, however, it is far from straightforwardly reducible to the category of an artwork either. Instead, the aim for this part is to position 'Machine with Concrete' as an exceptional technology.

To do this, I will focus on a series of tensions between simplicity and complexity that can be found across Ganson's general practice, and that are evident in this work in acute ways.[25] Assembled from a basic set of industry standard materials, 'Machine with Concrete' is nevertheless a work with a headily complex aim; indeed, presupposing consistency in the laws governing the physical universe, the ultimate aim of the work will prove impossible to meet. What renders 'Machine with Concrete' an exceptional technology, then, is that it is a work with a self-consciously impossible aim.[26] In turn, this renders the work a potentially instructive focus for considering issues of possibility, practice and environmental impact across contemporary fields of art, engineering and design.

Ganson arrived at the initial concept behind 'Machine with Concrete' in 1992, and there have been four main iterations since (Stern 2013: 229; Ganson 2009). The fourth version, from 2009, has the alternative title 'Beholding the Big Bang', but Ganson's recorded statements make it clear that he conceives of this piece as a refinement of his initial concept (Ganson 2009). In this part, I will focus mainly on the second version of the work, from 1992, and on 'Beholding the Big Bang'.

The second version of 'Machine with Concrete' (Figure 7) is perhaps the best known of all Ganson's works, with large numbers of viewers online (Ganson 2008). It consists of an electric motor connected to twelve worm drives arranged in series and mounted on a wooden base. Fully assembled, the work is less than one metre long. Here is how Ganson describes it:

> This machine was inspired by dreaming about gear ratios and considering the unexpected implications of exponential powers. ... Each worm/worm gear pair reduces the speed of the motor by 1/50th. Since there

FIGURE 7 *'Machine with Concrete'*.

are 12 pairs of gears, the final speed reduction is calculated by (1/50)12. The implications are quite large. With the motor turning around 200 revolutions per minute, it will take well over two trillion years before the final gear makes but one turn. Given the truth of this situation, it is possible to do anything at all with the final gear, even embed it in concrete. (2017)

What is most striking about this passage is the contrast between the directness of Ganson's tone and the implications of his concept. This, however, is of a piece with many other tensions opened up by the work. Most obviously, 'Machine with Concrete' plays on tensions between lived phenomenological time (the revolutions of the motor) and 'deep' time (the anticipated but never to be witnessed turning of the later gears) (Talasek 2014). However, the work also opens up tensions between our expectations concerning the functions and limitations of low-tech hardware, over and against emergent aesthetic qualities these might take on as parts of a whole (see Simondon 2012b). Further still, a marked tension between humour and seriousness is embodied in the work: is it to be written off as a self-consuming joke or gimmick, or should more sombre interpretations be projected into it?[27]

One of the most interesting tensions opened up by 'Machine with Concrete' concerns possibility and impossibility. Consider Ganson's final remark in the above passage: 'Given the truth of this situation, it is possible to do anything at all with the final gear, even embed it in concrete.' The 'given' here is that of the situation where the final gear makes 'but one turn', simply by following what the work is programmed to do. This situation is conceivable and programmable, and so possible in principle. By virtue of all manner of material and temporal conditions, however, it is inconceivable that this situation should ever be realized.

To draw out the headiness of this 'inconceivability', let me pursue some reflections on 'Machine with Concrete' here. The 'given' of this work, paradoxically, is that of an impossible situation, given as impossible. To draw out the implications of this point, consider merely the most obvious and immediate constraints that stand in the way of 'Machine with Concrete' actualizing its concept. Under normal circumstances, electric motors can stand several years of continuous functioning, but even this is on the condition that their bearings are regularly checked and changed. In commercial construction projects, concrete has a required design working life of from fifty to one hundred years (Cather and Marsh 1997: 26). In normal conditions of exposure at room temperature, the steel of the worms and screws used in the work may start to rust after several years, and the treated wood of its base will start to corrode after several centuries. Scaling up, any number of further constraints emerge to add complexity to this picture: from functional constraints like the wear and tear of torque on the screws, to ecological constraints like the presupposed availability of a steady power supply, to anthropological constraints concerning the longevity and stability of the human economies and civilizations providing the general conditions of the work's display.

Now consider the next part of Ganson's remark in the above passage: 'it is possible to do anything at all with the final gear'. Intentionally or not, this works like a materialization of the 'principle of explosion' from classical logic: 'from a contradiction, anything follows'. Once again, we witness a tension between humour and seriousness here: of all the possibilities that could conceivably have followed, embedding the final gear in concrete is both very modest, and one that plays on our received sense of the world's solidity in profound ways.

To draw this out, consider a tendency to take 'concreteness' and 'the concrete' as synonyms for that which is most solid, real and permanent, whether in everyday speech, or in philosophical language. Consider, for instance, the use of these terms in the existential tradition, or, indeed, in empirical turn philosophy of technology (Sartre 1960; Verbeek 2005: 7). Ganson's work playfully explodes this presupposition of synonymy, but his decision to embed the final gear in concrete should by no means be read as a dead end on possibilities opened up by 'Machine with Concrete'. On the contrary, the principle of explosion embodied by the work also opens more subtle but wide-ranging possibilities for modifying the work's concept.

Perhaps in an echo of the divisions of the Roman calendar, the second version of 'Machine with Concrete' has twelve worm/gear pairs. Combined, this results in a more or less arbitrary projected figure of over two trillion years before the final gear will turn.[28] With the 2009 work, 'Beholding the Big Bang', Ganson appears to invert this relation of precision and arbitrariness. For this work, a larger number of worm/gear pairs is assembled to project a more precisely calculated figure for the turning of the last gear:

> In remaking ['Machine with Concrete'], Ganson mused that life on earth is estimated to last only another billion years and that the earth itself will only last another 5 billion. 'The machine is going to be wasted,' he half jokes. 'So I thought 13.7 billion years is a nice significant number – that's the current estimated age of the universe since the Big Bang. I calculated this one so that the last gear turns once every 13.7 billion years.' (Harris and Lyon 2013)

Like the second iteration of 'Machine with Concrete', 'Beholding the Big Bang' is a work of deceptive simplicity. It is assembled with store-bought components, on a small scale (5 × 34 × 8 inches (Talasek 2014)), and all the tensions described above are once again evident in it (especially between humour and seriousness). There is, however, a further feature of this iteration that can wrong-foot the viewer who brings prior knowledge of 'Machine with Concrete' to it. On closer inspection, 'Beholding the Big Bang' turns out to have twenty-four worm/gear pairs, meaning that it also participates in the duodecimal ordering system embodied by the second iteration of

'Machine with Concrete', only less obviously so. This harbours a chastening joke: faced with a work dealing with orders of magnitude on an aeonic scale, average human perception can struggle to take in one increment of twelve on first impression.

But why have I bothered going into this level of interpretive detail on this work? Is it not simply self-indulgent or hermetic? Put simply, in what sense is 'Machine with Concrete' an exceptional technology with anything to say about the design, implementation and use of technologies today?

Consider, in the first instance, some resonances between 'Machine with Concrete' and contemporary work in the field of design. On the relation between design and possibility, the philosopher Mads Nygaard Folkmann writes:

> The ability to address, mediate, and evoke new possibilities, thereby creatively exploring new territories of use, meaning, and impact, is a defining feature of design. It is capable of transforming the possible into actual, tangible, useful objects that can have a huge impact on human life and behaviour (with widely distributed products) or on widespread notions of what objects are or mean (in experimental design). ... The possible is found not only before and after the realization of the design object but is also contained within it. (2013: 3)

'Machine with Concrete' is a work with an impossible aim. A first inclination might therefore be to position it as a particularly abstruse or ironic instance of experimental design, as referred to by Folkmann. There are, however, also notable parallels between the work and methods employed by commercially successful industrial and product design companies. Consider the practice of 'design fiction', as coined by the science fiction writer Bruce Sterling, and pursued by industrial design agencies such as 'Nonobject'. This practice involves imagining prototypes for superfluous and impossible artefacts, including a 'Superpractical Cell Phone' covered entirely in buttons, and kitchen furniture made from an undiscovered element (Lukíc 2011: 33, 121). In this sense, 'design fiction' is a method for charting the outer reaches of contemporary design from within the industry, and has been described effusively as an 'epistemological probe' making use of methods of 'humour', 'disruption' and 'extrapolation' by the design historian Barry M. Kātz (2011: xxv–xxvii).

'Machine with Concrete' is arguably a design fiction par excellence. This is because it resonates with each of the elements just identified by Kātz, and, moreover, because it does so from a position that sits *outside* the mainstream commercial design industry. First, 'Machine with Concrete' makes use of humour in profound tension with seriousness. Second, the work is disruptive: not necessarily in the sense of notions of 'disruptive technology' and 'disruptive innovation' current in contemporary economics and business studies, but because it instantiates qualities that disrupt *clichés* associated with these notions.[29] Third, in what 'Machine with Concrete' does with ideas of time and possibility, it both resonates with and restrains Kātz's hyperbolic description of 'extrapolation' as a process of 'turning an idea into a metaphysical absolute and extend[ing] it infinitely in every direction' (2011: xxvii).

From here, we can shift to consider some of the broader philosophical context surrounding work in design. As thinkers including Bruno Latour and Peter Sloterdijk have noted, the concept of 'design' has been extended considerably since the latter part of the twentieth century (Latour 2008; Sloterdijk 2005). Indeed, on Latour's reading, the concept should in fact be extended much further, to a point where it comes to be recognized as a better concept than that of 'revolution' for meeting challenges posed by current political and environmental crises (2008: 2). Latour's claim here, perhaps hyperbolically, is that design has the capacity to meet such crises in a problem-oriented way that falls outside 'heroic' theories of action privileged by what he calls 'modern' notions of the human subject (2008). With this in mind, consider what he has to say about 'details':

> [An] important implication of design is an attentiveness to details. ... A mad attention to the details has always been attached to the very definition of design skills. And 'skill' is actually a term that is also attached to design, in the same way that design is associated with the words 'art' and 'craft'. In addition to modesty, there is a sense of skilfulness, craftsmanship and an obsessive attention to detail that make up a key connotation of design. (2008: 3)

What I have sought to show over the course of this part, when read in light of these remarks, is twofold: first, that 'Machine with Concrete' is a work manifesting this type of attention to detail, and, second, that it merits sustained philosophical attention because of this.

Viewed in this way, 'Machine with Concrete' emerges as a nexus for issues affecting many different traditions in art and design: it is at once a manifestation of engineering, tinkering, invention, a 'found object' practice, 'serious play' (Sicart 2014), and, insofar as Ganson is a self-taught practitioner, a kind of 'outsider art'. Further still, it can also be read as a work of conceptual art, and, indeed, Ganson invites this kind of interpretation when describing his general practice of arriving at concepts before starting to work with materials (Harris and Lyon 2013: 36). As noted above, however, Ganson also describes his practice as one of 'gestural engineering' with links to kinetic sculpture, and this moves him closer to traditions of puppetry and theatre (Harris and Lyon 2013: 36; Blume 1998).

But doesn't accentuating this level of specificity simply imply further problems? Kinetic sculpture had a short-lived high point in the 1960s, before dispersing to inform movements like op art (Lee 2004). Is it not therefore backward-looking to foreground a work with links to this movement? Why not highlight older moments in the history of art, such as the surrealist approach to *'les machines célibataries'* (Carrouges 1976)? Given the general drift of art practice towards digital since the 1990s, why not aim for a wider-ranging study of more contemporary movements like net.art, glitch art, or 'post-internet' art (Chatonsky 2013)? Alternatively, why not a focus on a single work that connects to themes concerning the digital in more obvious ways, such Eva and Franco Mattes's 'Perpetual Self Dis/Infecting Machine' (2001), or Thomson and Craighead's 'Stutterer' (2014)?

My point in this part has not been to deny that other focuses are possible and desirable. Instead, it has simply been to show how sustained focus on a work like 'Machine with Concrete' can offer an entry point into a nexus of broader issues affecting practices of technology, art, engineering and design.

To draw out some of the stakes of this approach, consider the following challenge to designers, laid down by Latour at the end of the article just cited above:

> Here is the question I wish to raise to designers: where are the visualization tools that allow the contradictory and controversial nature of matters of concern to be represented? A common mistake (a very post-modernist one) is to believe that this goal will have been reached once the 'linear', 'objectified', and 'reified' modernist view has been scattered through multiple view points and heterogeneous make shift assemblages. ... What is needed instead are tools that capture what have always been the hidden practices of modernist innovations: objects have always been projects; matters of fact have always been matters of concern. The tools we need to grasp these hidden practices will teach us just as much as the old aesthetics of matters of fact – and then again much more. ... What I am pressing for is a means for drawing *things* together. (Latour 2008: 13. Original emphasis)

The suggestion I want to forward in concluding this part is that case studies of exceptional technologies such as 'Machine with Concrete' are one way to meet this challenge. This is because such case studies are designed to be ways 'drawing things together' in detail, while foregrounding explicitly paradoxical aspects that allow what Latour calls 'matters of concern' to emerge.

By 'matters of concern', Latour means issues such as globalization and the Anthropocene, that he takes to implicate human and non-human entities existentially, and not merely 'in theory' (Latour 2014).[30] With this in mind, consider how 'Machine with Concrete' can act as a form of 'visualization tool' for such matters. In addition to the issues already mentioned in this part, it can be read as:

1   A focus for issues to do with consumer culture and sustainability (is it a critique of the concept of 'planned obsolescence', conducted over an incredibly *longue durée*?)

2 A critique of conceptions of science and technology considered as heroic 'modernist' pursuits based around the 'eureka' moments of solitary geniuses (is the work some form of Arthurian 'sword in the stone', constantly provoking its viewers to try to solve its puzzle by manually turning the last gear?)
3 An endorsement of public and non-profit education projects in art, science and technology (when we consider the types of contexts in which 'Machine with Concrete' and 'Beholding the Big Bang' have been displayed to date).[31]

These are possible interpretations, not necessary ones. Insofar as it harbours these possibilities, however, the case study conducted in this part can be framed as just the type of 'visualization tool' that Latour called for above: 'Machine with Concrete' is not an assemblage designed to scatter attention, but something that stills, slows and focuses it; by having a 'project' that is impossible, the work challenges our sense of what constitutes an object in profound ways, and connects to multiple 'matters of concern'; by having resonances with the sublime and movements in the history of art, the work can appear reducible to what Latour above called 'old aesthetics' – given a different type of attention, however, it in fact emerges to trouble this category, as a manifestation of a singular practice sitting somewhere between the contexts of art, technology and design.

4 Problems and prospects

In this chapter, I have focused on a merely imagined technology, a failed technological practice and a technology with an impossible aim. As outlined at the beginning, my aim has not been to offer a definitive set of exceptional technologies, or to wrest these case studies from other disciplines where they are often covered in much more comprehensive detail. Instead, my aim has been to suggest some more content for the concept of exceptional technologies, and to highlight the potential for links across a variety of disciplines and fields, in favour of an expansive sense of what philosophy of technology can aspire to today.

What needs to be emphasized in this respect is that each of the case studies conducted in this chapter has very clear and criticizable limits. For instance, each case study relates to an artefact or practice produced by a lone male inventor, and each might be described as 'hobbyist' in different ways. My hope, however, is that these limitations are contingent to the cases studied, and not essential to the concept of exceptional technologies: instead of depriving the concept of value, my hope is that they in fact set out a very clear agenda for how it be might criticized, clarified and developed further.[32]

Something else to be emphasized in this respect (and to be returned to in the conclusion of this book) is that a focus on 'exceptional technologies' does not imply a form of technological 'exceptionalism'. Exceptionalism involves the fetishistic claim that some entity or class of entities is *a priori* better than all others across all possible contexts in some vague sense. The claim underpinning this chapter is very different. It is this: for any given context of technological design, implementation and use, there will be artefacts and practices that show up as marginal or exceptional to its presuppositions and norms, and these may be just as useful for drawing out and focusing on the presuppositions and norms of the context in question. Indeed, in a general contemporary context where our sense of technologies and their implications can seem to be accelerating, complexifying and specializing away from us at great rates, the claim is that such exceptional technologies can sometimes be more valuable for focusing and nuancing our sense of such conditions. In the conclusion to this book, I will come back to these issues.

CHAPTER FIVE

Which way to turn?

The previous chapter sought to give entry points for the concept of exceptional technologies, and concluded by considering how this concept might be developed further. But what about the other key terms of this book's title? Throughout, I have considered continental philosophy in a way that associates it with Kant and an open sense of the transcendental focused on conditions, and I have focused mainly on the empirical turn when discussing philosophy of technology. But isn't this doubly out of date? For the past decade or so, continental philosophy has been undergoing a putative 'speculative turn' that has, it is claimed, been taking it away from the supposedly moribund 'correlationist' or 'anti-realist' inclinations of Kant and the transcendental, towards a proliferation of speculative and 'weird' realisms (Bryant et al. 2011; Sparrow 2014: 115; Braver 2007). Similarly, it might be objected that the 'empirical turn', occurring in the late 1990s and early 2000s, is by now a dated matter in philosophy of technology.

When considered in light of the particular conjunction between continental philosophy and philosophy of technology that this book has sought, these problems might only seem to get worse. If continental philosophy has taken a turn in a speculative direction and philosophy of technology has taken a turn in an empirical direction, doesn't this mean they have turned away from one another radically: the former towards the 'abstract', the latter towards the 'concrete'? The worst-case scenario, it seems, is this: not merely have I overlooked continental philosophy's speculative turn, and not merely have I overstated philosophy of technology's empirical turn; more fundamentally, I may seem to have overestimated the extent to which continental philosophy and philosophy of technology can be viewed as vehicles on the same road – instead, they may have turned off in radically opposed directions.

The aim of this chapter is to dispel this worst-case scenario in favour of better articulating the approach to method involved in this book. To do this, it focuses on a shared picture of turning that can be viewed as underpinning and conditioning the speculative and empirical turns, in spite of their otherwise considerable differences. On one level, this picture may appear trivial. The argument of this chapter, however, is that it is in fact fundamentally important, because it is a crude picture: for every turn towards, according to it, attention must turn away from something else, and this, I will argue, is problematic because it works against openness to new entities and problems emerging to challenge the limits of our purview in philosophy of technology, and because it blocks potentials for cross-disciplinary work.

To be clear, my aim in challenging this picture of turning is not to suggest that most work in either recent philosophy of technology or continental philosophy reduces to it. Instead, it is to point out how inadequate the picture is for describing the work that has gone on in these fields to date, as well as how divisive it can become when viewed as a prescription for the type of work that can go on in and between them in the future.

There are, in this sense, two related questions motivating this chapter: How can philosophy of technology turn towards the empirical without turning away from more speculative 'transcendental' issues concerning conditions for the empirical (as exhibited by work across the continental tradition, and not merely since its putative speculative turn)? And, how can we turn towards such speculative issues without turning away from specificity, detail and focus (as traditionally aspired to by empirically focused philosophy of technology)? The argument is that what may be blocking us from engaging with these questions, at least in part, is an underlying and inadequate picture of 'turning' as method. What is incumbent upon us, I argue, is to challenge this picture in favour of alternative pictures of method that enable the emergence of a more thoroughgoing sense of philosophy of technology as a shared field in relation to which the empirical and speculative turns discussed above are taking place.

Part 1 considers how the picture of turning just discussed has played out in recent philosophy of technology. On the one hand, it is correct to view the empirical turn as a dated issue. The argument of this part, however, is that its influence is in fact ongoing on a more fundamental level in philosophy of technology in terms of the precedent it has set for turning as a key logic of innovation. This builds upon and extends remarks I made towards the end of Chapter 1 of this book.

Part 2 considers how the picture of turning has played out in different ways in recent continental philosophy. On the one hand, it is plausible to view the speculative turn as a radical break in the history of this field, and this impression is strengthened by the bombast that recent speculative approaches typically direct against the continental tradition since Kant. The argument this part, however, is that certain recent speculative approaches in fact mark a complexification of a sense of the transcendental manifest across the broader continental tradition, and that this can be admitted without losing what might be philosophically provocative or interesting about these approaches.

Part 3 considers the dangers and merits of an alternative picture of method as 'mapping' for an approach to philosophy of technology that would be open to speculative and empirical developments alike. I develop this through a focus on Foucault's famous account of the 'panopticon' in *Discipline and Punish*, which I read as an exemplary attempt to map an 'exceptional technology'.

Part 4 concludes the chapter by considering how the picture of mapping discussed in Part 3 might better enable us to engage complexities involved in doing philosophy of technology today, in contrast to the pictures of turning discussed in Parts 1 and 2.

1 The empirical turn: An enduring influence in philosophy of technology?

In 1998, Peter Kroes and Anthonie Meijers hosted a conference at the University of Delft that forwarded a programmatic call for an empirical turn in philosophy of technology. In their introduction to the subsequent collection of papers, they positioned the empirical turn as an attempt to offer a 'reorientation of the field', and described philosophy of technology as a 'discipline in search of its identity' (2000a: xvii). According to their vision, the empirical turn was to imply the following:

> Philosophy of technology should keep its distinctive philosophical nature. Nevertheless, it should also base its analyses on empirical material, much more than has been done so far. ... The philosophy of technology should concentrate more on the clarification of basic conceptual frameworks used in the engineering sciences and in the empirical sciences studying technology and less on abstract myths and fictions of which it is not clear how they relate to the real world of technology. ... These [empirically-focused] descriptions regard technology as it is conceived in engineering practice, as well as technology as it functions in our everyday world. (2000a: xxi)

Towards the end of Chapter 1, I focused on three problematic issues concerning the empirical turn. First, I claimed that, although empirical turn philosophers are right to criticize a tendency to reify 'Technology' in certain 'classical' continental approaches to philosophy of technology, they tend to perpetuate the same mistake against the broader continental tradition when it comes to the theme of the 'transcendental'. Second, I claimed that the empirical turn is committed to a sense of what constitutes a 'Technology' that leaves too much to common sense. Third, I claimed that the empirical turn has laid down a problematic precedent for innovation in philosophy of technology through commitment to a picture of turning as method.

The key aim for this part is to show how the third of these issues has played out in philosophy of technology since Kroes and Meijers made their call for an empirical turn. Since this concerns how all three of the issues just reviewed interrelate, however, let me first make some brief remarks on the other two.

Although Kroes and Meijers's call for more thoroughgoing analyses based on 'empirical material' is laudable, it is noteworthy that they make this point in opposition to work 'done so far' in philosophy of technology. Further remarks make it clear that, by this, they mean 'metaphysical analyses of technology (under the influence of Heidegger)', and 'critical reflections on the consequences of science and technology for the individual and social form of life' (Kroes and Meijers 2000a: xvii). In this respect, Kroes and Meijers are consistent with critics of insufficiently empirical 'transcendental' approaches, as discussed in Chapter 1 (Verbeek 2005; Achterhuis 2001). As I noted above, there is certainly something in this criticism in specific cases (perhaps especially that of Heidegger). What this does not provide, however, is sufficient warrant for reifying 'transcendental' approaches *en bloc*, in order to write them off as irrelevant to work in philosophy of technology. On the contrary, on the account developed in this book, doing so in fact stigmatizes a valuable set of resources for meeting the aim of a more thoroughgoing and dynamic engagement with 'empirical material' in philosophy of technology.

Now consider Kroes and Meijers's reference to 'abstract myths and fictions'. Perhaps they have in mind the types of metaphysical and critical analyses just discussed here, or perhaps something more folkloric still, of the order of urban myths concerning technology. The key issue, however, is this: Kroes and Meijers assert that such 'abstract myths and fictions' have 'no clear relation to the real world of technology'. What is problematic about this statement is that it presupposes a relation in order to deny it. Put simply, it seems that Kroes and Meijers can only claim that 'abstract myths and fictions' are not clearly related to the 'real world of technology' on the strength of a very clear sense of what actually constitutes the relation between these two sets of entities.[1] In this sense, their statement has less to do with 'myths and fictions', and more to do with 'the real world of technology': along with their reference to 'technology as it functions in our everyday world', it seems intended to strengthen a rhetorical impression that this 'real world' exists as an unproblematic and relatively stable object of common sense.[2]

According to the key claim to be developed in this chapter, the points that Kroes and Meijers make on these two issues are related to an underlying picture of turning as method. Before moving to consider how broader work in philosophy of technology relates to this, let's first observe how Kroes and Meijers view this picture:

Our call for an empirical turn in the philosophy of technology is ... not to be understood as a wish to turn this branch of philosophy into an empirical science, nor as a plea to turn it away from normative/evaluative matters. ... It is a call to base philosophical analysis concerning technology on reliable and empirically adequate descriptions of technology (and its effects). In our opinion, this is a *condition sine qua non* for the philosophy of technology to be taken seriously in present-day discussions about technology. This does not mean that its primary focus should be on empirical problems, for that would turn it into an empirical science. Its focus should be on conceptual problems, more in particular, on the clarification of basic concepts and conceptual frameworks employed in empirically adequate descriptions of parts or aspects of technology. (2000a: xxiv)

This passage exhibits a nuanced consideration of some of the implications that follow from buying into a picture of turning as method. Kroes and Meijers make it clear that, by 'turning' they do not mean metamorphosis into an empirical science (a 'turning into'), nor rejection of normative/ evaluative matters (a 'turning away from'). Taken together, these two points give the impression that what philosophy of technology should be most wary of turning away from concerns ethics and politics. However, there are crucial epistemological and ontological issues that this understates, and that will have knock-on effects for any subsequent conception of the types of ethical and political issues to be discussed: put simply, what is to count as the object of what Kroes and Meijers describe as an 'empirically adequate description of parts or aspects of technology', and what is to count as 'an adequate description'?

The broader issue here is this: insofar as philosophy of technology is not turned towards questions like this in a thoroughgoing way, but instead reliant on an appeal to some form of common sense to settle them, then there is a great deal more that it stands to turn away from. Considered merely in terms of Kroes and Meijers's above account, for instance, it is implied that philosophy of technology should turn away from 'metaphysical analyses of technology', 'critical reflections on the consequences of science and technology for the individual and social form of life', and a (vaguely defined) set of 'abstract myths and fictions'.[3] As Kroes and Meijers note, it may be that such issues are 'not clearly related' to a common-sense conception of 'the real world of technology'. But 'no clear relation' does not mean 'no relation' at all. On the contrary, if the task of philosophy is, as Kroes and Meijers state, one of 'clarification of basic concepts and conceptual frameworks', it may instead mark out the issues just discussed as excellent candidates for further investigation, in favour of a more dynamic and expansive conception of philosophy of technology.

While Kroes and Meijers were aware of issues affecting their picture of turning, then, they were wrong to take a gesture towards 'normative/evaluative' issues to be sufficient to redress this. Nevertheless, their approach can be viewed in terms of a precedent that the empirical turn set for subsequent work in philosophy of technology more broadly. As Peter-Paul Verbeek puts it:

> In retrospect, one could say that in recent decades the philosophy of technology underwent first an 'empirical turn' and then an 'ethical turn'. … The descriptivist orientation that resulted from the empirical turn was compensated for in the first decade of the twenty-first century, which saw an explosion of ethical approaches to technology. A broad variety of ethical subfields emerged, including nanoethics, ethics of information technology, ethics of biotechnology, ethics of engineering design, and more. This rapid growth of applied ethical approaches to technology can partly be seen as the result of the empirical turn. Rather than criticizing 'Technology' – as classical philosophers of technology often did, pointing out its potential threat to 'humanity' – ethical reflection started to address actual technologies and technological developments. (2011: 160–2)

According to this summary, subsequent work in philosophy of technology has tended to play out in terms of Kroes and Meijers's initial concerns about their picture of turning: aware of the potential dangers of 'descriptivism', philosophy of technology has, on Verbeek's account, tended to construe an appeal to ethics as sufficient 'compensation' for all that the empirical turn stood to turn away from. As discussed above, however, a turn towards ethics cannot constitute such a form of compensation if neither the 'empirical' nor the 'ethical' turns involve a more thoroughgoing attempt to address ontological and epistemological issues concerning what constitutes a technology or what constitutes an adequate description of a technology.

In making this point, I am not claiming that these issues are avoided by all (or even by most) of the approaches that might conceivably be grouped under Verbeek's 'empirical' and 'ethical' turn headings. Instead, I am claiming that a picture of turning as method is inadequate to describe the approaches that do not avoid them, and that it prescribes a misleading norm for those that do. To draw this out, consider a tension in the position Verbeek himself goes on to develop. On the one hand, Verbeek is aware of the insufficiency of a gesture towards ethics if all this implies is a form of 'externalism', where pre-existing ethical theories get indifferently applied to case studies. Drawing inspiration from Bruno Latour's approach to science and technology studies (STS), Verbeek's preferred position is one of 'moral mediation', where ethical theories emerge as 'coshaped' by technologies that are recognized to act as moral agents or 'actants', and not as lumps of morally indifferent 'matter' (Verbeek 2011: 163). Having made this point, however, Verbeek writes:

> In order to see this interwoven character [of morality and technology], we need to integrate the empirical-philosophical approach to technology with normative reflection. We need to make – ... with a nod to Latour – one more turn after the empirical turn. (2011: 164)

This passage indicates a great deal about how 'locked in' to a picture of turning as method some approaches in philosophy of technology may have become in the wake of the empirical turn. Following Latour, Verbeek's aim is to draw attention to the sense in which 'coshaping' occurs in rich and case-specific ways (2011: 163). The point, however, is that Verbeek's recourse to a notion of 'one more turn' is deeply counterproductive for making this point: rather than allowing his approach to stand up for itself, it makes it seem like one more variation on an underlying methodological theme that philosophy of technology has adopted since the empirical turn.[4]

The broader point here is this: a picture of turning may have become so engrained in philosophy of technology that even thinkers as reflexive as Verbeek opt for it as a default logic of innovation, even when it is inadequate for describing what their approaches are really after. To this, it could of course be objected that we have just taken Verbeek's case in isolation. Let me therefore put it in the context of a point I made in Chapter 1 of this book, concerning the proliferation of other calls for 'turns' forwarded in the recent history of philosophy of technology. Since Kroes and Meijers made their call for the empirical turn, there have, for instance, been calls for an 'engineering turn' (Vermaas 2016), 'ethical' or 'axiological' turns (Kroes and Meijers 2016), a 'societal turn' (Brey 2016), a 'semantic turn' (Krippendorff 2005), a 'practice turn' (Hillerbrand and Roeser 2016), a 'narrative turn' (Kaplan 2009b), a 'policy turn' (Briggle 2016), as well as recent calls to emulate anthropology's 'ontological turn' (Descola 2013).

Again, the aim in highlighting this profusion of turns is by no means to suggest that work conducted under their headings has nothing to say. Instead, it is to suggest that a picture of turning as method is inadequate and divisive for getting their respective messages across. This is because, although this picture may appear trivial in fact, it commits work in philosophy of technology, in principle, to an approach that carries connotations that are crudely first person, voluntarist, oppositional and progressivist. Where it is our dominant picture of method, the nagging sense will therefore be that, for any turn in philosophy of technology to be possible, it will have to aim, in principle, at 'winning the field'. What's more, this is likely to remain our implicit sense of things where an underlying picture of turning is uncritically accepted, even if, in fact, the respective turns occurring on the field make explicit moves to appear more conciliatory than this.

What this picture of method ignores (as discussed towards the end of Chapter 1 of this book) is that all the respective turns just mentioned imply a sense of philosophy of technology as a shared field. In this respect, any move occurring on the field implies all the others, more or less directly or indirectly. Instead of a rich sense of how all this plays out at different levels of complexity, in terms of potentials for conceptual innovations, case studies and crossover work with other fields, the tendency of philosophy of technology's picture of turning has, however, unfortunately been towards a sense of ever more specialized, fractured and exclusive turns. Furthermore, this is likely to remain the case, in principle, where our underlying picture of method is one of turning, even if it constitutes a gross misrepresentation of the kind of work that philosophers of technology are, in fact, seeking to do.

In retrospect, the empirical turn was but one turn occurring in the field of philosophy of technology, and can be viewed as a dated matter when considered in terms of other turns to have occurred since. Insofar as it set the precedent for the logic according to which these other turns have occurred, however, the influence of the empirical turn is ongoing on a more fundamental level. This influence is skewed, and a gesture towards ethical issues is not sufficient to redress it. What is needed instead, I suggest, is a sense of the different possibilities that alternative pictures of method might open up for philosophy of technology as a field.

2 The speculative turn: A new beginning in continental philosophy?

Before moving to consider a specific example of an alternative picture of method in the next part, let me use this part to consider how a picture of turning relates to recent developments in continental philosophy. Although this picture has recently been applied to make sense of moves to push continental philosophy in a putatively more 'speculative' direction, I will argue that it is even less adequate as a description or prescription for work in this context than in philosophy of technology. The reason for this is that, while these 'speculative' developments can be viewed as a radical turn away from Kant's legacy for the continental philosophical tradition on a superficial level, they can also be viewed as a complexification of a sense of the transcendental manifest across this tradition on a more fundamental level.

In the introduction to their 2011 collection, *The Speculative Turn: Continental Materialism and Realism*, Levi Bryant, Nick Srnicek and Graham Harman give this summary of their concept of a speculative turn:

> The speculative turn ... is not an outright rejection of ... critical advances [made since Kant]; instead, it comes from a recognition of their inherent limitations. Speculation in this sense aims at something 'beyond' the critical and linguistic turns. As such, it recuperates the pre-critical sense of 'speculation' as a concern with the Absolute, while also taking into account the undeniable progress that is due to the labour of critique.

The works collected here are a speculative wager on the possible returns from a renewed attention to reality itself. In the face of the ecological crisis, the march forward of neuroscience, the increasingly splintered interpretations of basic physics, and the ongoing breach of the divide between human and machine, there is a growing sense that previous philosophies are incapable of confronting these elements. ... The new turn towards realism and materialism within continental philosophy comes in the wake of something resembling ethereal idealism. (2011: 3)

As discussed in Chapter 1 of this book, a key impetus for what Bryant et al. describe as the speculative turn here came with the publication of Quentin Meillassoux's *Après la finitude* in 2006. After this, two important conferences in the field took place in 2007 and 2009, under the titles of 'Speculative Realism' and 'Speculative Realism/ Speculative Materialism'. All five of the key speakers across these events contribute chapters to *The Speculative Turn* (Meillassoux, Graham Harman, Ray Brassier, Iain Hamilton Grant and Alberto Toscano). Additionally, there are chapters in the collection from other thinkers associated with the speculative realist theme (including Bryant himself, Adrian Johnson and Steven Shaviro), as well as contributions from key figures working on themes of realism and materialism in the continental philosophical tradition more broadly (including Isabella Stengers, Manuel DeLanda, Slavoj Žižek, Bruno Latour, Francois Laruelle and Alain Badiou).

Bryant et al. recognize from the outset of their collection that 'it is difficult to find a single name adequate to cover all [the] trends' to which their assembled thinkers relate. Nevertheless, they 'propose "The Speculative Turn"' as a title, reasoning that it provides 'a deliberate counterpoint to the now tiresome "Linguistic Turn"' (2011: 1). In this part, I want to emphasize that this title is even less adequate than Bryant et al. suspect. This is not merely because it is insufficient to cover the considerable differences between their assembled thinkers.[5] More fundamentally, the issue concerns the relation between the implied picture of 'turning' and Kant's legacy for the continental tradition.

Consider Bryant et al.'s reference in the above passage to the 'inherent limitations' of critical advances made in philosophy since Kant. What is at the root of these is the anthropocentrism of Kant's transcendental idealism. As discussed in Chapter 1 of this book, transcendental idealism is the metaphysical doctrine according to which fixed *a priori* categories inherent to the subject construct reality as it is given to the subject. Notoriously, this commits transcendental idealism to the position that the subject can only know reality as it appears, and that reality 'in itself' can never be known. Since publication of *Après la finitude*, this position has been referred to as 'correlationism' in speculative circles, and its influence has been taken to extend well beyond the specifics of Kant's approach. As intimated above, Bryant et al. view the 'linguistic turn' as a permutation of correlationism, to the extent that it takes reality to be constructed by language; however, the general tendency in the speculative literature has been to extend correlationism's legacy much further, to the point where it emerges as the dominant postulate behind most of the main forms of post-Kantian continental thought – from German idealism, hermeneutics and critical theory, to structuralism, post-structuralism and phenomenology (Brassier 2007: 50; Sparrow 2014: 86–93). When Bryant et al. refer to 'ethereal idealism' towards the end of the above passage, it is the playing out of correlationism's legacy that they have in mind. According to this picture, the post-Kantian continental tradition exhausted the various 'critical' and 'linguistic' turns of correlationism and ended up stuck in a free-floating play of text and discourse that turned away from pressing issues concerning the 'material' and the 'real' (see also James 2012).

Viewed in terms of this picture, Bryant et al.'s proposed 'speculative turn' emerges looking like a radical break indeed. What it aims to turn away from is nothing less than the Kant-inspired legacy of correlationism for a continental tradition that emerges looking moribund as a whole. What it aims to turn towards, on the other hand, are forms of realism that will seem 'weird' or outlandish according to the biases of any anthropocentric philosophical perspective. As Bryant et al. put it:

> The various strands of continental materialism and realism are all entirely at odds with so-called 'naïve realism.' One of the key features of the Speculative Turn is precisely that the move toward realism is not a move towards the stuffy limitations of common sense, but quite often a turn towards the downright bizarre. (2011: 7)

This is the aim of what Bryant et al. call the 'wager' involved in recent speculative approaches: to break the deadlock of correlationism in favour of 'weird realisms' that challenge more 'naïve' forms of common-sense realism. The reason such common-sense realism should be challenged, in turn, on Bryant et al.'s account, is because they take it to be consistent with the anthropocentric biases of correlationism: rather than challenging common-sense presuppositions on what can legitimately appear as 'real' to the subject, naïve realisms are precisely 'naïve' on Bryant et al.'s account to the degree that they expect the real to appear in a consistent 'more of the same' fashion that renounces the possibility of a real beyond the limits of common sense, capable of intervening to challenge these limits. As Harman puts it in attempting to define speculative realism:

> By 'realist' I mean that these philosophies all reject the central teaching of Kant's Copernican Revolution, which turns philosophy into a meditation on human finitude and forbids it from discussing reality in itself. By 'speculative' I mean that none of them merely defend a dull commonsense realism of genuine trees and billiard balls existing outside

the mind, but a darker form of weird realism bearing little resemblance to the presuppositions of everyday life. (Harman 2010: 2)

Let's dwell on the terms of this summary for a moment, in order to give a focused sense of what may be faulty with the general picture of a 'speculative turn'. First and foremost, the central teaching of Kant's philosophy, according to the position advanced over the course of this book, is not, as Harman puts it, that philosophy should be turned into 'a meditation on human finitude [that] forbids it from discussing reality in itself'. Instead, it is that philosophy involves a sense of the transcendental as an approach to argument or method: given X, an approach is 'transcendental' where it enquires into conditions for the possibility of X. As developed in Chapter 1, this sense of the transcendental is not reducible to the terms of Kant's transcendental idealism. Instead, what 'transcendental' connotes, on this reading, is an ongoing inquiry into conditions that is applicable to a range of different metaphysical approaches. What accounts for the differences between these approaches, in turn, is how far they go towards problematizing and radicalizing the key terms of inquiry: What is the 'given'? To whom or what is it given? What is withheld and not 'given'? In what does 'inquiry' consist? What is a condition? What is conditioned ('X' or 'objecthood')?

The central teaching of Kant's approach, on this reading, is not metaphysical, but methodological.[6] While this shift of emphasis may appear slight, it in fact has important consequences for how we picture developments in recent continental philosophy. The key point is this: while speculative approaches aiming at realism and materialism can be viewed as discontinuous with what they position as the idealist and anthropocentric metaphysics of 'correlationism', they are continuous on a more fundamental level with a sense of the transcendental as an approach to argument or method.

For this reason, a picture of 'turning' is inadequate as a description for recent speculative approaches. This is because it frames them, *en bloc*, in terms of a unilateral 'turning away' from a preceding continental tradition that it caricatures as moribund. Further, the picture of turning also emerges as a divisive prescription for approaches seeking to build on recent speculative work in continental philosophy. This is because it makes it appear as if these approaches are generated ex nihilo and voluntaristically, when in fact they emerge as complex permutations on an underlying approach to argument or method that is prevalent across the continental tradition in philosophy more broadly.

Neither of these points implies that recent speculative approaches have nothing interesting or novel to say. On the contrary, it allows us to account for their novelty in terms of the challenging and paradoxical permutations they introduce into a sense of the transcendental: what speculative approaches provocatively take as their 'given' are the limitations of correlationism itself, as evidenced in the traditional literature, and what they attempt to account for, through speculative moves, are conditions for the possibility of this given that are irreducible to its conventionally anthropocentric, idealist and common-sense biases. Contrary to 'common sense' and 'everyday' presuppositions on what constitutes the 'real', we therefore get a profusion of speculative approaches such as Brassier's 'eliminative nihilism', which contends that Nature is the radically indifferent and intrinsically meaningless condition for thought and experience (2011: 48), Meillassoux's 'speculative materialism', which contends that the only necessary condition of thought and experience is that all of their conditions are utterly contingent (2006: 107–08), Harman's radically non-relational 'object-oriented' approach, which, with a nod to Kant's concept of the 'thing in itself', holds that no object whatsoever can ever be fully known or experienced, as a matter of metaphysical principle (2010: 105–21), and Bryant's 'onto-cartography', which, with a nod to Deleuze and Guattari, undertakes to redescribe entities as 'machines' (2014). These positions may appear absurd according to the limits of correlationism; in order to appear so, however, they take a sense of the transcendental beyond correlationism.

On a superficial level, then, the speculative developments discussed in this part can be viewed as a radically new beginning for continental philosophy, insofar as they undertake to challenge the metaphysics of correlationism. This impression is strengthened by rhetorical tendencies in the speculative literature towards bombast, which give the impression of a general 'turning away' from what they position as a moribund continental tradition, as well as the impression of a new and emerging movement turned towards correlationism's 'beyond' (what Meillassoux calls its 'great outdoors' (2006: 37)). Insofar as these speculative developments can also be viewed as permutations and complexifications on a sense of the transcendental, however, they are more fundamentally continuous with developments across the continental tradition more broadly.

For these reasons, the picture of a 'speculative turn' is more inadequate than we might initially suspect. To this, it might of course be objected that I have simply read too much into a contingent and trivial word choice by Bryant et al. over the course of this part. However, these issues of triviality and contingency are precisely to the point. Bryant et al.'s choice of 'turning' is less trivial in terms of its implications than they suspect (because it commits them to a picture of method that has crudely first person, dialectical, voluntaristic, oppositional and progressivist implications). Crucially, however, this picture is also more contingent than Bryant et al. suspect (because alternative pictures of method, as I will seek to show in the next part, are both possible and desirable). What I take the notion of a 'speculative turn' to show, in this sense, is not just how inadequate, but also how *contingent* applications of the picture of turning as method can be. With this in mind, let me now conclude with a consideration of how an alternative picture of method might work for philosophy of technology today.

3 An alternative picture: Method as 'Mapping'

Picture a series of interactive and evolving maps related to a space, on which it is possible to zoom in and out in terms of complexity, detail and abstraction, and on which it is possible to chart the emergence of new entities, problems and connections. Imagine these maps are not just synchronically related, but diachronically related, in that they allow for changes in the space to be charted over different temporal registers (whether, for instance, geographical, political, economic or social). Imagine also that they have topological functionality: it is possible to simplify their elements in order to draw out relations between other maps and the elements on them. Imagine, crucially, that the limits of these maps are apparent: rather than being maps that could easily be mistaken for their territory, they are contingent ways of making sense of the space in which they are situated, in relation to which they emerge, and which exceeds them on all sides.

This, I submit, is an alternative picture of method to which philosophy of technology might productively aspire today: as 'mapping'. Against what has become something of a default impulse for work in the field, we would have to be wary of positioning this picture itself in terms of a picture of turning (as if it constituted some form of 'cartographical' turn). If desired, pictures of turning could of course be retained in philosophy of technology, because the aim is not so much to replace or turn away from them, as to *emplace* them in terms of their conditions of possibility: that is, to make turning less of a default impulse for work in the field, and to make it more of a reflexive move, occurring in terms of a broader and more open sense of the space in which it is situated.

This picture would of course carry its own dangers. Most obviously, it could be taken to imply a fantasy of total knowledge involving a 'God's Eye' overview or 'view from nowhere' (see Nagel 1986). In practice, however, it might constitute a significant improvement on a picture of turning in precisely this respect. This is because, whereas the notion of a theoretical 'turn' seems to imply a crudely first-person perspective related to a highly abstract sense of space, a picture of mapping implies a sense of perspective that is already collaborative and technologically mediated. Indefinitely many others (both human and non-human) are required to produce the texture of a map, both in terms of its referents and its coordinates. What's more, rather than being omniscient, the kind of perspective that requires the technological mediation of a map is one that must recognize its own limits (whether implicitly or explicitly) relative to the situation in which it is embedded.

Suppose a world in which technologies are operative to be the space that 'mapping' in philosophy of technology sets out to chart. What might it aspire to? Consider the following five points:

1 It can aspire to move further in both empirical and transcendental directions at once: rather than turning towards 'technologies' on the strength of a relatively fixed conception of what constitutes an object of actual or possible experience, it can strive to chart empirical changes in technologies as well as changes in the conditions under which sense gets made of and through these entities (whether, for instance, in a political, economic, aesthetic or cognitive sense).

2 It can aspire to operate at high levels of both specificity and generality, at once: rather than turning towards localized case studies to the exclusion of more global issues, and vice versa, philosophy of technology can aspire to case studies that are 'scalable', and that connect both to specialized issues and to issues that might seem to lie beyond the immediate purview of philosophy of technology (for instance: the functioning of the eighteenth-century Newcomen steam engine in relation to the theme of the Anthropocene (see Kroes 2000: 28–40; Stiegler 2015)).[7]

3 It can aspire to go further in drawing together diverse thinkers and philosophical traditions: rather than treating areas of its map as 'no go zones' based on the acquired habits of other branches of philosophy as historically evolved, philosophy of technology can aspire to draw more extensively on different thinkers and traditions to better address the diversity of issues raised by its case studies. This can be done by noting points of contact and divergence between thinkers in analytic, continental and non-Western philosophical traditions and by applying methods and concepts to case studies that productively draw out points of contact and difference (see, for instance, Hui 2016).

4 It can aspire to chart points of both overlap and divergence with a broad range of other disciplines, from the engineering sciences and philosophy of science, through to design, art practice, software studies, cultural studies, cognitive science and more.

5 It can aspire to chart a field where points of contact and divergence between the various traditional branches of philosophy get foregrounded in terms of issues concerning practice and material culture: rather than turning towards ethics in order to redress the perceived imbalances of an 'empirical turn', philosophy of technology can chart a field where issues raised in ontology/

metaphysics, epistemology, logic, ethics, and aesthetics clash and become focused, with consequences for philosophical work more broadly (on this, see Pitt 2016; Cassirer 2012).

Instead of being exhaustive, this list is intended to be suggestive, open-ended and open to critique. It may, moreover, be that other pictures of method stand to offer a better sense of the type of work that might be carried out in philosophy of technology today.[8] The broader point, in this sense, is simply this: we should be willing to seek out the limits of our underlying pictures of method, and to vary them, in favour of opening further vectors for work in the field.[9]

This point takes us back to concerns outlined at the beginning of this chapter. To close, let me now attempt to address it with reference to a specific example, which I will then relate back to the pictures of turning discussed in Parts 1 and 2. The example I have in mind is Foucault's treatment of the panopticon in *Discipline and Punish*.

On the account developed in this book, Foucault's approach to the panopticon is an exemplary attempt to map an 'exceptional technology'. By this, I mean that it develops a highly focused and specific account of an artefact that might otherwise appear as marginal or exceptional to our received sense of what constitutes a technology, and, in the process, raises challenges that have the capacity to recalibrate how we make sense of technologies and their implications more broadly. In doing so, Foucault's approach can, I think, be shown to move in both empirical and transcendental directions at once, to demonstrate high levels of both specificity and generality, to draw on a range of disciplines beyond the purview of philosophy 'proper', and to nevertheless raise important issues for philosophy as traditionally conceived. Crucially, his approach is speculative (because it extends the implications of the panopticon well beyond the initial context of its emergence), yet has empirical purchase (because it does not aim to cover all possible contexts, and because the limits of its application are apparent and criticizable).

The panopticon was, famously, the utilitarian philosopher Jeremy Bentham's 'simple idea' for an architectural installation formed of a central observation tower surrounded by a ring of cells, to be used for the correction, inspection and training of individuals in the interests of society, 'no matter how different, or even opposite the purpose' (Bentham 1995: 34). It was, according to Bentham, to be an 'engine' for '… *punishing the incorrigible, guarding the insane, reforming the vicious, confining the suspected, employing the idle, maintaining the helpless, curing the sick, instructing the willing* in any branch of industry, or *training the rising race* in the path of *education*' (1995: 34, Original emphasis). In Foucault's account, the panopticon is read both in terms of Bentham's literal description of a 'particular institution, closed in upon itself', and as a 'figure of political technology that may and must be detached from any specific use' (1991: 205). This enables Foucault to read it both as a new kind of 'seeing machine' (1991: 207), and as a figure for the emergence of what he calls 'disciplinary society' or 'panopticism' over the course of the eighteenth and nineteenth centuries.

The panopticon is liable to appear as marginal or exceptional to our received sense of technology in a number of ways that the subsequent notoriety and influence of Foucault's account may have served to obscure. Most obviously, it was a merely imagined building.[10] Read critically, this might immediately make us wonder whether or not Foucault's account is simply playing games with metaphors, and stretching our concept of a 'Technology' well beyond the scope of its legitimate application. Read more charitably, however, what Foucault may in fact be doing is this: introducing paradoxical challenges for our sense of what constitutes a technology in order to extend this concept beyond the habitual scope of its application.

Famously, Bentham envisaged the panopticon's central tower being fitted with shutters and partitions that would prevent inmates in the surrounding cells from being able to verify whether or not they were being observed at any given moment (Bentham 1995: 35–7). As Foucault puts it:

> The Panopticon is a machine for dissociating the see/being seen dyad: in the peripheric ring, one is totally seen, without ever seeing; in the central tower, one sees everything without ever being seen. ... A real subjection is born mechanically from a fictitious relation. (1991: 201–2)

This short passage is at once highly specific to the panopticon, yet broad in its speculative implications. On the level of the specific, Foucault draws out the sense in which the panopticon was to serve as a direct material intervention into the process of vision itself. On a naïve phenomenological level, to see another is, in general, to presuppose the possibility of being seen by that other (Sartre 2003: 276–26). As envisaged by Bentham, however, the panopticon was to break this reciprocity: through a direct material intervention, it was intended to constrain what all those subject to it (both inmates and observers alike) could count as objects of their actual or possible experience.

On a more speculative level, this has the capacity to provoke reflection on other forms of technological mediation that either break or impose similar reciprocities: from seemingly trivial cases like one-way mirrors, through to politically charged cases concerning hacking, surveillance, the use of satellites and warfare conducted with drones (see, for instance, Chamayou 2013). The speculative implications of Foucault's remarks do not stop there, however. Consider his closing statement on 'real subjection'. What this points to is the capacity of technologies to induce and mediate forms of 'placebo effect': just as it is possible for the panopticon to produce 'a real subjection' from a 'fictitious relation', so too is it possible to get lost in a book, to become attached to an internet avatar, or to react somatically to a piece of music or film.[11]

To be sure, these examples are much more mundane than that of the panopticon. But this, I suggest, may be precisely what makes the latter an important instance of an 'exceptional technology'. By developing a clear and specific focus that challenges our sense of what constitutes a technology, Foucault's approach to the panopticon points in both empirical and speculative directions at once: by inviting us to consider empirical features of the panopticon as envisaged by Bentham, it provokes further wide-ranging reflection back on other more ostensibly 'everyday' artefacts and practices.

Foucault's treatment of the panopticon as a figure for 'disciplinary society' or 'panopticism' should, I suggest, be positioned at the limit of this process. On Foucault's account, an 'functional inversion' took place across Western societies in the eighteenth century, where institutions that had previously served as marginal sites for the excluded under the conditions of what he calls 'sovereign society' began to emerge as centres of 'techniques for making useful individuals' (1991: 211).[12] What recommends the panopticon as an appropriate figure of this shift, in turn, is that its position in accepted discourse mirrors this historical change. As Foucault writes:

> It must be recognized that, compared with the mining industries, the emerging chemical industries or methods of national accountancy, compared with the blast furnaces or the steam engine, panopticism has received little attention. It is regarded as not much more than a bizarre little utopia, a perverse dream. (1991: 224–5)

What authorizes Foucault's shift from a focus on the empirical specificity of Bentham's 'bizarre little utopia' to the speculative generality involved in reading 'disciplinary society' in terms of a form of 'panopticism' here? On one level: historical changes from the marginal to the central. On another level: the fact that drawing attention to the case of the panopticon structurally mirrors these changes at the level of discourse. At his most bombastic, this leads Foucault to declare: 'In appearance, [the panopticon] is merely the solution to a technical problem; but, through it, a whole type of society emerges' (1991: 216). This statement could easily be dismissed as an instance of crude technological determinism. What this would overlook, however, is precisely the duality of Foucault's approach. Empirically, the panopticon *was* 'merely the solution to a technical problem', and Foucault's approach invites us to consider it as such. On a more speculative level, however, his approach also provokes us to consider the panopticon as a focal point or 'object lesson' in the wide-ranging forces that can go into constituting a society: from its 'technical solutions' to its 'bizarre little utopias' and 'perverse dreams' alike.

These passages are, I hope, sufficient to suggest something of the sense in which Foucault's approach to the panopticon involves a complex attempt at 'mapping'. While this picture may not be adequate to capture all of the complexities involved in his approach, any temptation to describe Foucault's approach in terms of a facile conception of a 'turn' (towards 'discourse analysis', for instance) would, I think, be even less adequate.[13] As developed above, this is because Foucault's approach works on multiple different levels and directions at once. Besides involving high levels of both specificity and generality, it

1. involves both empirical and transcendental issues (insofar as it describes the panopticon as a new object of sense, and provokes us to consider it as a new figure for permutations in the conditions under which sense gets made of objects more broadly);
2. draws on a range of disciplines beyond philosophy 'proper' (witness Foucault's broader use of historical, criminological and journalistic sources throughout *Discipline and Punish*);[14]
3. raises important issues for the branches of philosophy as traditionally conceived (for instance: to what extent can the panopticon be placed in terms of the ontological/metaphysical, epistemological, logical, aesthetic and ethical commitments of Bentham's utilitarianism more broadly?).[15]

4 A shared field of exceptional complexities

How does the picture just discussed relate back to the pictures of turning discussed in Parts 1 and 2 of this chapter? As developed in Part 1, a key danger facing philosophy of technology since the empirical turn may consist in its tendency towards ever more fragmented turns, based on a narrow conception of what constitutes a 'technology'. What Foucault's approach exemplifies, in contrast to this, is how issues in philosophy of technology might be approached with attention to both the specific and the general, and to the empirical and the speculative, at once.

The case of the speculative approaches discussed in Part 2 is in some ways more complex. Certainly, many of these approaches touch on issues in philosophy of technology in interesting ways. A key danger they raise, however, consists in the temptation to turn speculation into a kind of 'free phantasy' (Zahavi 2016: 304), as if deranging the referents and coordinates of our conceptual maps in favour of a proliferation of new ontologies could be considered sufficient to dispense with the task of remapping our understanding. Consider, for instance, Graham Harman's 'object-oriented' approach, which develops a provocative reading of Heidegger:

> Although Heidegger speaks of 'tools', and although most of his examples are such things as hammers and railway platforms, the tool-analysis actually holds good for anything. All objects are encountered more often as tacit components of our world than as blatant objects of

awareness. This is just as true of people, numbers, and religious edifices as of shovels and drills. Heidegger's tool-analysis is actually a universal theory of entities, although this point is still missed by the majority of commentators. (2011: 110)

There is a suggestive speculative move at the heart of these remarks. In *Being and Time*, Heidegger famously reads the absence or breakdown of a tool (its 'unreadiness to hand') as a provocation for *Dasein* to re-examine its implicit understanding of the world, considered in terms of the holistic network of relations that give the tool its meaning and function (Heidegger 2005: 91–148). On Harman's reading, however, this Heideggerian approach remains too anthropocentric and holistic. For Harman, a tool's 'unreadiness to hand' should more fundamentally be read as a sign of its radical independence as an entity, not merely from any of the meanings and functions that *Dasein* might assign to it, but, further, from reduction to any relation whatsoever into which it might enter (2011: 105–21).

In principle, there may be nothing to stop this kind of provocative speculative move leading to focused and interesting work. But what has followed in fact? Consider Harman's above reference to 'people, numbers … religious edifices …, shovels and drills'. Rhetorical tropes like this can be found throughout Harman's work: rather than offering focused case studies, his approach has, unfortunately, tended towards a kind of literary 'maximalism', where surreal conjunctions of entities proliferate. As Nathan Brown puts it:

> [Harman] leaves us with no meaningful criterion for the constitution of objects at all. Should any of this seem unpersuasive, we are submitted to the rhetorical coup de force of constantly reiterated allusions to parrots and glaciers and quarks, etc., etc. Since all kinds of objects are mentioned, this really must be a philosophy of objects. Distraction is what passes for epistemology. (2013: 63)

Perhaps with this kind of problem in mind, Levi Bryant has developed a speculative account that crosses over with issues in philosophy of technology, and that develops a picture of 'mapping' as method. He defines his approach like this:

> 'Onto-cartography' – from 'onto' meaning 'thing' and 'cartography' meaning 'map' – is my name for a map of relations between machines that analyzes how these assemblages organize the movement, development, and becoming [of] other machines in a world. (2014: 7)

What, then, is a 'machine' on Bryant's account? Drawing inspiration Harman's speculative reading of 'tool-being', and from Deleuze and

Guattari's approach to what they call 'desiring machines' in *Anti-Oedipus*, Bryant writes:

> Being has never consisted of anything but machines. ... 'Machine' is ... our name for any entity, material or immaterial, corporeal or incorporeal, that exists. 'Entity', 'object', 'existent', 'substance', 'body', and 'thing', are all synonyms of 'machines'. (2014: 15)

Elsewhere in his book, Bryant writes that 'the paradoxical danger of cartographical analyses is that they can blind us to ... new forms. ... We forget that the map is not the territory and that ... territories change' (2014: 284–5). In Bryant's own approach, the 'map', it seems, is 'onto-cartography', understood as the practice of mapping relations between 'machines', and the 'territory' is being itself, understood as 'machinic' all the way down. One problem with this picture, however, is that Bryant's own claim that 'being has never consisted of anything but machines' would seem to involve a fairly radical instance of mistaking the map for the territory. Of itself, this kind of speculative move might still prove provocative and suggestive.[16] There is, however, a deeper problem: Bryant's speculative move can give the impression that problems of specificity have been addressed, when in fact the register for addressing them has simply been shifted.

The deep problem facing the type of speculative move that Bryant's 'onto-cartography' is premised on is not that it is too controversial or absurd. Instead, it is that it can be trivially admitted without really being informative, because it does not dispense with the hard work of drawing attention to how such a change of perspective impacts on specific cases in nontrivial ways. As with Harman, however, Bryant's approach unfortunately emerges as deeply distracted on this issue: instead of a focused consideration that makes good on the initial provocation of his speculative move, what emerges from his approach is a proliferating set of allusions and more or less trivial redescriptions.[17]

When Foucault approaches the panopticon in *Discipline and Punish*, I suggest, he does so in a way that avoids these tendencies in favour of a focused epistemology that incorporates empirical and speculative elements alike. His approach might therefore seem less spectacular than recent speculative turn approaches because it avoids their kind of self-consciously 'weird' ontological gambits. What Foucault's might really have avoided in this respect, however, is a tendency to quickly become the prisoner of its own rhetoric, in favour of allowing more interesting and focused speculative issues to emerge in less forced ways.[18] In the process, I suggest, Foucault's approach also appears avoid a different pitfall raised at another extreme of concerns in philosophy of technology: that of dismissing the panopticon as a 'non-entity' on the basis of a restrictive empirical turn purview, because of its status as an artefact that was merely imagined (a 'bizarre little utopia').

For these reasons, I suggest that Foucault's approach to the panopticon constitutes an exemplary attempt at mapping the complex implications of an exceptional technology. Making this point does not, I submit, amount to some form of dilution of empirical concerns in philosophy of technology, nor to some form of reactionary turn against more recent speculative approaches in continental philosophy. Instead, it is to point out that theoretical 'turns' will inevitably be misdirected when they ignore the exceptional complexities that go into constituting the fields on which they are taking place. What Foucault's approach exemplifies, in contrast, is how a focus on an exceptional technology can act as a way of drawing out these complexities.

Conclusion: Exceptional technologies, not technological exceptionalism

This book has forwarded a concept of 'exceptional technologies' for work in philosophy of technology and related fields. In doing so, it has presupposed that we can aspire to a broad and dynamic picture of what work in philosophy of technology can look like today, and that we should not prejudge the range of fields to which this work might be related in interesting ways. Over the course of the book, three main premises have underpinned my approach: 1) that philosophy of technology stands to benefit from a more thoroughgoing engagement with methods, concepts and thinkers drawn from the continental tradition in philosophy; 2) that a concept of exceptional technologies is a justifiable addition to the conceptual 'map' or 'toolkit' of philosophy of technology today; and 3) that important and timely issues concerning material culture and practice are at stake in philosophy of technology, and that our capacity for meeting these can be developed further through a consideration of our pictures of method.

In concluding, I want to emphasize a specific point: the approach forwarded in this book does not amount to a form of what might be called 'technological exceptionalism'. 'Exceptionalism' tends towards the position that some entity or class of entities is to be privileged across all possible contexts.[1] This book has not presupposed that exceptional technologies form such a class of entities. Instead, it has presupposed a different sense of the exception: as that which is marginal, paradoxical or excluded according to a received sense of things in a given context. The argument of this book has been that, for any given context of technological design, implementation or use, there will be such exceptions to what constitutes our received sense of a technology, and that these exceptions can often be deeply instructive for focusing on problems and issues affecting the received sense, and for challenging its limits.[2]

At the limit of technological exceptionalism, we find the tendency to treat a reified and homogeneous notion of 'Technology' as a privileged way of making sense of contemporary reality as a whole. As I have argued in this book, approaches in philosophy of technology since the empirical turn have been right to criticize this tendency, and they have been right to identify it in some key approaches emerging from the continental tradition. As I have also argued, however, we should be wary of repeating this mistake, by reifying a sense of the transcendental evident in the continental tradition into some form of otherworldly 'Transcendental' realm, and by homogenizing continental approaches to philosophy of technology in terms of a bygone 'classical' approach. On the contrary, I have argued that a renewed sense of the transcendental as an approach to argument or method in fact stands to make profound sense for philosophy of technology today: first, as a way of better engaging with methods, concepts and thinkers in the continental tradition; second, and relatedly, as a way of tracking changes in the conditions that constitute our received sense of a technology across different contexts, on different levels of complexity: from the micro to the macro, and in terms of empirical and speculative issues alike.[3]

How does this sense of the transcendental relate to the concept of exceptional technologies? Viewed critically, this book's attempts to outline this concept might easily seem to exercise an apparent vice of recent continental philosophy: towards a creation of concepts that threatens to multiply entities well beyond necessity.[4] Put simply, can't what I have called 'exceptional technologies' be adequately made sense of through a host of other current concepts: from 'hobbyism', 'pseudoscience', 'science fiction' or 'futurism', through to 'design fictions', 'new media' or Bruno Latour's notion of a 'quasi-object'? (Latour 1993: 73–4). Isn't Vannevar Bush's memex simply an amateurish 'thought experiment'? Isn't Francis Galton's composite photography simply pseudoscientific 'propaganda'? Isn't Arthur Ganson's *'Machine with Concrete'* just a dated work of 'kinetic sculpture'?

By 'exceptional technologies', I mean artefacts and practices that have the capacity to focus a sense of the transcendental on conditions that might otherwise be overlooked or excluded in a given context of technological design, implementation or use. Of itself, a sense of the transcendental merely opens the way to this concept and does not justify it. What justifies the concept, I take it, are the *preconceptions* that we might associate with terms like those just listed above. Put simply, I take it that preconceptions attached to these concepts might cause us to overlook or split apart specificities of artefacts and practices that a new concept of exceptional technologies has relatively better capacity for calling to attention.[5] For this reason, I take the concept of exceptional technologies to be a justifiable addition to philosophy of technology's conceptual 'map' or 'toolkit', while fully acknowledging that the elaboration of it undertaken in this book has been partial and far from exhaustive.

CONCLUSION

If this book has not been aiming at a form of 'technological exceptionalism', then why has it focused on philosophy of technology at all? Doesn't this very choice of focus betray a secret form of 'exceptionalism'? On the contrary, and as I outlined in the introduction, my key premise for focusing on philosophy of technology has simply been this: it is a field where issues concerning material culture and practice are encountered in important and timely ways, and it is a field where potentials for dynamic and heterodox forms of philosophical thinking are at stake.

My hope is that the approach developed over the course of this book as a whole has been able to sufficiently demonstrate this premise. In Chapter 1, I argued for a renewed sense of the transcendental, not as a metaphysical 'realm' out of touch with the empirical, but as an approach to argument or method that can be dynamically attuned to empirical concerns. While the continental tradition in philosophy is by no means reducible to the theme of the transcendental, I argued that it can act as an important guiding thread for drawing out key strengths of this tradition for philosophy of technology today. In Chapter 2, I considered how an ostensibly trivial technology (the blank page) can be viewed as 'exceptional' in the sense discussed above. In doing so, I argued for a renewed sense of Edmund Husserl's 'imaginative variation' as a materially and historically situated practice. Viewed in this way, imaginative variation emerges not as a vague form of idealist 'fantasy' but as a way of enacting and focusing a sense of the transcendental in practice, and of blocking its theoretical tendencies towards infinite regress. In Chapter 3, I focused on embodiment conditions to show how a sense of the transcendental is already operative across a range of contemporary approaches that draw on the continental philosophical tradition: from an existential/phenomenological approach to the internet, through contemporary media theory, to work in '4e' cognitive science.

Chapter 4 then offered three more focused case studies of exceptional technologies, as a way developing this concept further, and of showing how it might be related to a range of different fields, including software studies, data visualization, design and art practice. I focused on cases of merely imagined, failed and impossible exceptional technologies, but concluded by emphasizing ways in which the concept might be criticized and developed further. After this, Chapter 5 considered the limits of an underlying picture of 'turning' as method in philosophy of technology. Although this picture can appear trivial, I argued that it may in fact have fundamental consequences for how we conceive of work in the field, and that it should be challenged in favour of other pictures. To draw this out, I attempted to develop an alternative picture of method: as 'mapping'.

It is common for philosophy of technology to be referred to as a 'field', and it is common for philosophers of technology to discuss 'turns' taking place in this field. In order to challenge more culturally pervasive pictures of 'Technology' as a kind of road on which we are stuck, however, it may be necessary to engage in a more thoroughgoing consideration of what our underlying pictures of method imply when thinking philosophically about technologies, in order to make sense of their implications in new and dynamic ways. The claim advanced in this book is that such a process should always be possible, because any settled picture of what constitutes a technology will always have significant exceptions.

Notes

Introduction

1 For some very different perspectives, see Lanier (2013), Carr (2015), Virilio (2006), Fukuyama (2002), Stiegler (2015), Gleick (2002) and Gere (2008).
2 See the transhumanism of Bostrom (2005) or the accelerationism of Mackay and Avanessian (2014), and Srnicek and Williams (2016).
3 See Taylor (2014).
4 See Badiou (1999: 46–52).
5 See, for instance, Cowan (1990) and Morozov (2013) on 'technological solutionism'.
6 To paraphrase a notorious comment from Heidegger (2010), see also Henry (2003).
7 This book takes the continental tradition in European philosophy as a key influence, and crosses over with work in the analytic philosophical tradition, and with work in fields including media theory, cognitive science, and art and design. The book's engagements with philosophical positions outside the Western tradition are unfortunately constrained and sporadic. Rather than disqualifying the approach of the book as a whole, my hope is that this simply sets out a series of clear trajectories for future work aimed at developing and criticizing its approach. For an excellent recent account of philosophy of technology in the context of the Chinese philosophical tradition, see Hui (2016).
8 These have included neo-Heideggerian approaches (Borgmann 1984; Scharff 2010), new 'critical theory' approaches (Feenberg 2002, 2005), approaches drawing on McLuhan (Van Den Eede 2013), approaches related to the Actor-Network Theory of Latour (Latour 1992b, 2005; Harman 2002, 2007), the philosophy of information of Floridi (2013, 2014), approaches drawing on philosophy of language (Coeckelbergh 2017) and the 'pharmacological' approach of Stiegler (1998, 2015). For surveys of general trends, see Berg Olsen et al. (2007, 2009), and Scharff and Dusek (2011).
9 The concept of the empirical turn emerged from a series of focused studies by Dutch philosophers of science, technology and engineering at the turn of the millennium (see, in particular, Kroes and Meijers 2000b, Achterhuis

2001). These observed signs of an empirical turn in diffuse late-twentieth-century approaches to philosophy of technology, in a way that exceeded the specifically Dutch context. In Kroes and Meijers (2000b), signs of this are detected in the Social Construction of Technological Systems (SCOT) approach (Bijker, Hughes and Pinch 1987), Kuhn's paradigm approach to philosophy of science (1962), the 'Strong Programme' in sociology of science of the Edinburgh School (Bloor 1976) and the science and technology studies approach of Latour (2005). In Achterhuis (2001), signs are detected in the work of six contemporary North American philosophers of technology: Albert Borgmann, Hubert Dreyfus, Andrew Feenberg, Donna Haraway, Don Ihde and Langdon Winner. It is arguable that Kroes and Meijers (2000b) and Achterhuis (2001) present two different conceptions of the 'empirical turn'. The key point, however, is that, in both cases, the 'empirical turn' is construed as a paradigm shift in philosophy of technology at the turn of the millennium, and not as a local event in the Dutch or the North American context. For reflections on the continued relevance of the empirical turn, see Aydin and Verbeek (2015) and Franssen et al. (2016b).

10 Consider discussions of 'roadmaps' in contemporary contexts of design, business, innovation and warfare/reconstruction (NSA 2003). Consider also discussions of the 'drivers' of technological innovation (Roe Smith and Marx 1994), the concept of 'path dependence' in technological development (Page 2006) or the concept of the 'pacing problem' in legal studies (Marchant et al. 2011).

11 Etymologically, a paradox is that which runs 'alongside' or 'counter to' (*para*) received opinion or common sense (*doxa*) (OED 2017). On the potential instructiveness of paradoxes in a logical sense, see Priest and Berto on 'Dialetheism' (2013).

12 Reviewing these examples, it might be objected that what the concept of 'exceptional technologies' aims to cover is already well-covered by concepts like 'pseudoscience', 'science fiction' or 'hobbyism'. However, such concepts are, I think, too well-worn and divisive. Using them in the case of the three examples I have just given, for instance, the tendency would be to separate these into distinct categories, and to treat them as *clichéd*. For similar reasons, I think the concept of 'new media' is inadequate to capture what is at stake in the case of more contemporary examples of exceptional technologies (on this, see the engagement with Mark B. N. Hansen's concepts of 'new media' and 'twenty-first-century media' in Part 2 of Chapter 4). Let me also take this opportunity to distinguish the concept of exceptional technologies from the concept of the 'state of the exception' that Giorgio Agamben has picked up and developed from the work of Carl Schmitt: the sense of 'exception' implied by this legal/political concept is distinct from the one developed in this book (see Agamben 2005).

13 Other exceptional technologies may include, for instance, technologies yet to be invented, thought experiments (Kuhn 1981; Dennett 2013), forgotten technologies (i.e. artefacts whose conditions of use and legibility have disappeared), waste and abandoned technologies (Chatonsky 2013) and, perhaps pre-eminently, artworks (Carrouges 1976).

14 Trivially, the claim of this book on this issue might be summed up like this: 'transcendental philosophy is worthwhile philosophy too!' I would be quite

happy for this claim to be recognized as trivial, because it seems to be what the contemporary turns discussed above dispute (along with large sections of the analytic philosophical tradition – see Chapter 1). Another way of stating things is that what this book is concerned with is a sense of the transcendental as a 'metaphilosophical' problem. On the importance of the transcendental for continental philosophy, see Chase and Reynolds (2011); Deleuze (2004a); Rockmore (2006).

15 Because it begins from an unclarified common-sense position on what constitutes a technology, the empirical turn runs the risk of being pushed and pulled towards a limited positivism, towards artefacts and practices privileged by marketing and towards what I call 'zeitgeist technologies'. On this, see Chapter 1.

Chapter one

1 These points recur throughout this book, and are picked up again in depth in Chapter 5.
2 See also Chase and Reynolds for an excellent account of the importance of transcendental reasoning in continental philosophy. As they put it, 'In the continental traditions starting with Kant but enduring throughout the twentieth century and beyond, some form of transcendental reasoning is close to ubiquitous' (2011: 89).
3 On attitudes to history of philosophy in the analytic tradition in philosophy, see Dummett (1996).
4 The reader discussed here is a straw man. I am consciously indulging this fallacy to exorcize a more troubling straw man concerning who or what the 'continental philosopher' is, and what they do. On analytic problems with the reception of continental philosophy, see Moore (2012: xx) and Williamson (2007: 286–7).
5 Quentin Meillassoux is the speculative turn thinker Malabou devotes most attention to in *Avant demain*. Her use of the term 'correlationism' is directly inspired by his (see Malabou 2014: 221–65). Meillassoux characterizes correlationism like this:

> By 'correlation', we understand the idea according to which we only have access to the correlation between thought and being, and never to one of these terms taken in isolation. We will therefore from now on call every current of thought which supports the unsurpassable character of the correlation understood in this way '*correlationism*'. (Meillassoux 2006: 18, My translation, Original emphasis)

6 In fact, Malabou's own approach in *Avant demain* turns out to involve such a non-dogmatic and dynamic sense of the transcendental (see Malabou 2014: 303–20).
7 Gardner identifies five key features of a separable 'transcendental turn'. I take his use of 'turn' here to be misleading but contingent, and undertake to offer a corrective to it through Malpas's emphasis on circularity in the next part (see Gardner 2015: 2–3; see also Chase and Reynolds 2011).

8 Chase and Reynolds take this abstraction of argument form to be definitive for the analytic reception of 'transcendental arguments', the genealogy of which they follow Stern and Stroud in tracing back to Austin, then developing via Strawson, then Stroud. In much work on transcendental arguments in the analytic tradition, references to Kant tend to be made by way of preface and historical contextualization (see Chase and Reynolds 2011; Malpas 1997: 3–4; Strawson 1966).

9 Chapter 3 returns to these issues concerning embodiment conditions in depth.

10 As most famously exemplified by the arguments offered in famous passages of the *Critique of Pure Reason*, such as the 'Transcendental Deduction of the Categories' and the 'Refutation of Idealism'.

11 On the characterization of Heidegger as an idealist, see also Lafont (2007: 106) and Moore (2012).

12 It was eventually published in 1804 (see Kant 2002).

13 As I will argue in Chapter 5, this point on conditions is compatible with recent speculative approaches that challenge the priority of experience in philosophy.

14 Chase and Reynolds raise such a suspicion (2011: 91).

15 This illusion is a 'transcendental illusion' in Kant's sense. Whereas Kant identified the illusory tendency to reify God, the soul, the universe, I am identifying the tendency to reify the transcendental method itself as a species of transcendental illusion (see Kant 2000).

16 These questions are important for meeting the suspicion that I have simply restated the principle of sufficient reason ('nothing comes from nothing'), and returned to a form of pre-Critical rationalism. This is because they trouble the presupposition that effects are contained, *a priori* and 'prospectively', in their causes; rather, they emphasize an initially regressive move from conditioned to conditions, in terms of relations that are not restricted to causality (such as those of difference, similarity, temporality and analogy, for example).

17 A further issue concerns the extent to which this minimal sense of the transcendental remains 'correlationist'. The answer is no, because it does not logically exclude the possibility of objects that are not and cannot be given acting as conditions for the given (on this, see Brassier 2011: 48–9). The approach developed in this part is directly inspired by Deleuze's 'transcendental empiricism' (see Deleuze 2004a; Sauvagnargues 2008; Bryant 2008; Reynolds 2014).

18 This part focuses on the empirical turn as an entry point for considering philosophy of technology today. For a justification of this, see Part 2 of this book's introduction.

19 For evidence of this tendency, see, for instance: Achterhuis (2001): 3, Brey (2008: 19–21, 2010: 1), Verbeek (2005, 2011), Ihde (1990: 159, 2010), Feenberg (1999: 183, 2002: 9, 2009), Michael (2006: 154), Cohen (2006: 145–7) and Aydin and Verbeek (2015). There are important exceptions to this homogenization. What of the treatment of technology in thinkers like Adorno, Lukács, Cassirer and Benjamin? Are their approaches 'classical' or not? (See Feenberg 2016).

20 Technological determinism is the thesis that technology '[induces] certain societal effects with necessity' (Fuchs 2011: 113), or, stronger, 'that technology

causes or determines the structure of the rest of society and culture' (Dusek 2006: 84). 'Autonomous Technology' is the related but distinct thesis that technology develops according to a logic that is autonomous from the rest of society and culture (Dusek 2006).

21 On this, see Ihde's influential concept of 'multistability' (2012). See also Verbeek (2005: 99–119) and Zweir, Blok and Lemmens (2016).

22 It must be emphasized that philosophy of technology already very much engages with continental philosophy, especially in its 'postphenomenological' and 'critical theory of technology' variants (see Verbeek 2005; Ihde 2012; Feenberg 2002). My point here is that shifting attention towards the theme of the transcendental stands to engage the continental tradition in a broader way still, by drawing in conceptual resources from thinkers for whom technology does not necessarily feature as an explicit theme (e.g. German idealists and a broader set of French poststructuralists, as well as some of the speculative approaches discussed in Chapter 5 of this book).

23 Examples of such technologies might include drones, genetic modification techniques such as CRISPR, cognitive enhancers, smartphones, ICTs, nanotechnologies and green technologies. I do not intend the term 'zeitgeist-seizing' to be dismissive or pejorative here, and my point is certainly not that philosophy of technology should avoid studying such artefacts. The point is that, while a focus on zeitgeist-seizing technologies is necessary for philosophy of technology, it is not sufficient, and should be supplemented with openness to 'exceptional technologies'. On related issues, see Edgerton (2006).

24 It should be noted here that historical case studies have always been a focus of authors such as Ihde (2012) and Pitt (2011), and are explicitly called for in recent reflections back on the empirical turn by Franssen et al. (2016a: 3). On materialist approaches, see Feenberg (2002) and Fuchs (2011).

25 In terms of history of philosophy concerns, I would like to note that this does not amount to a privileging of Hegel over Marx. Instead, it is to point out that Marxist and Hegelian approaches share the minimal sense of transcendental as an approach to argument or method that I have argued for above.

26 Stated differently: exceptional technologies are artefacts and practices capable of focusing a sense of the transcendental on conditions in a series of surprising and complex ways.

27 This is related to, but distinct from, Heidegger's sense of the 'Topology of Being' (see Malpas 2007).

Chapter two

1 A great deal of work in fields such as media theory, software studies and literary theory is influenced by a sense of the transcendental, and by continental philosophy more broadly (see, e.g. Bolter and Grusin 1999; Chun 2011; Hansen 2004; Galloway 2004, 2012; Hayles 2012). Chapters 3 and 4 of this book develop this in depth.

2   Locke famously writes:

> Let us then suppose the mind to be, as we say, white paper, void of all characters, without any *ideas*. How comes it to be furnished? Whence comes it by that vast store which the busy and boundless fancy of man has painted on it with an almost endless variety? Whence has it all the materials of reason and knowledge? To this I answer, in one word, from *experience*. In that all our knowledge is founded; and from that it ultimately derives itself. (1993: 45, Original emphasis)

'Remediation' is Bolter and Grusin's term for the process where new media assimilate the contents of older media (1999: 4–5). The point I am making here is that new media can also remediate old metaphors as media, such as that of the 'blank page'.

3   'Humanities philosophy of technology' is Mitcham's term for philosophy of technology that is continuous with the social sciences and humanities, rather than STEM subjects (Mitcham 1994; see also Franssen et al. 2013). It should also be noted here that this chapter in no way privileges the terms 'white' or 'blank'. Instead, my aim is to track conditions obscured by historically evolved figures of speech like tabula rasa, 'white paper' and 'blank page'. The broader point of this chapter is that 'white paper' and 'blank pages' are never 'white' or 'blank' in any absolute sense; instead, they are always conditioned in multiple complex ways that should be recognized and celebrated (see Derrida 1972).

4   For example, transcendental subjectivity in Kant's case (2000: 416–17), 'Absolute Spirit' or '*Geist*' in Hegel's case (1977) and, as examined below, a reworked form of transcendental subjectivity in that of Husserl (1970).

5   Husserl is, after Kant, the thinker in the continental tradition who is most regularly associated with the theme of the 'transcendental', and, from around 1905, he explicitly framed his phenomenology as 'transcendental' (Bell 1991). The rationale for focusing on *Ideas* and *The Crisis* concerns the influence these two works have had for the subsequent continental tradition: Figures as diverse as Sartre, Ortega, Scheler, Heidegger, Merleau-Ponty, de Beauvoir and Habermas were all directly influenced by them, and the concept of the 'Lifeworld' articulated in *The Crisis* is now common currency in fields as varied as philosophy, psychology, science and technology studies, and cultural studies. Furthermore, it is possible to locate the putative break between analytic and continental approaches in the 'transcendental phenomenology' developed in these books (see Chase and Reynolds 2011; Dummett 1996; Bell 1991).

6   On Husserl's concept of imaginative 'free variation', see Husserl (1997: 373–6, 2002: 134–7). See also Casey (1976: 56–8), Bell (1991: 194–5) and Moran (2000: 154).

7   On the use of imaginative examples in phenomenology, Casey writes: 'Phenomenological method as conceived by Husserl takes its beginnings not from uninterpreted givens but from carefully selected examples' (1976: 23).

8   When *Ideas* was first published in 1913, this presupposition was a contingent necessity. It is only since the advent of ICTs that Husserl's presupposition has been revealed to be even more contingent than this.

9   Husserl composed by hand, either at a sitting or standing desk, and completed his manuscripts by typewriter (Ihde 2016: 59–76).

10 It is also instructive to situate Bernard Stiegler's approach in this way. In Stiegler's case, this occurs through the exploration of 'technics' as an 'unthought' condition for the possibility of philosophy, and through engagement with Husserl's writings in terms of this theme, most notably in the first and third volumes of the *Technics and Time* series (Stiegler 1998: ix; see also Stiegler 2004: 14–15).

11 As mentioned above, I place no privilege on 'blank' or 'white' here. As will emerge over the remainder of this part and the next part, there is no such thing as a purely 'blank' or 'white' page, and this is why the complex conditions concealed by these apparently trivial figures of speech need to be examined (see n.3 above).

12 It should be noted here that blank pages are implicated in Husserl's writings in a further complex sense: as a recognized theme in the history of philosophy. In the published version of *The Crisis*, for instance, there are five key references to Locke's tabula rasa, three of which refer to Locke's 'white paper', using the English term in the original. This suggests a technical and by no means throwaway significance for this concept in Husserl's writings (Husserl 1970: 63, 85, 88, 89, 115). For a contrasting account of the tabula rasa, see Pinker (2002).

13 It might be objected that this characterization only applies to a select group of students (e.g. humanities students), to students in 'developed' nations (e.g. the US, Europe, Japan and South Korea), or that it is anachronistic (since conventional written exams may well be phased out by computer-based literacy). These are important considerations, but they are tangential to the purpose of the example, which is simply to differentiate between Husserl's situation and that of the hypothesized student.

14 Perhaps the closest comparable background condition we can imagine in Husserl's case concerns the role of libraries, whether personal or public, or the voluminous research notes Husserl kept. The point I want to make here is that removal of either of these conditions by exam conditions differs considerably in degree from removal of internet access.

15 Husserl died in 1938, and wrote in the late nineteenth and early twentieth centuries. For him, it would make sense to use the metaphor of the blank page as Locke did to discuss the tabula rasa, and even to discuss technological artefacts such as chalkboards, writing slates, artist's canvases or photographic film. Husserl would, however, have had no concept of the ways in which technological change has subsequently extended the metaphor to the innovations of computing.

16 My point here is not that masking complexity is the arbitrary aesthetic choice of Big Tech companies like Apple or Google. It is that such a process of 'masking' is a necessary logical condition on any interaction with new media, given the scale of the complexities they involve. Given this necessary condition, interfaces can then be further manipulated for all manner of contingent aesthetic, ethical and political ends (see Chun 2011: 1; Galloway 2012: 78–100).

17 Battelle comments: 'Nearly fifty per cent of all searches use two or three words, and twenty per cent use just one. Just five per cent of all searches use more than six words' (2005: 27).

18 Consider, for instance, Ihde's search for explicit examples of technologies in Husserl, as noted in Part 1 above.

19 One such inexhaustive list might include everything from 'objects', 'tools', concepts and traits of a style, to memories, distractions, beliefs, fears, moods, prejudices, nervous ticks, sensory *qualia*, pains and habits, to the entire weight of cultural, economic, political, technological and biological history, past, present and future.

20 Deleuze often slides towards all out polemic against 'presuppositions', 'preconceptions' and 'metaphors', as if to suggest that they had no utility for our thinking (see Deleuze 2004a; Deleuze and Guattari 2004a,b). Against this, his work on Bergson is a good point of contrast, developing Bergson's more tempered account of 'presuppositions', 'preconceptions' and 'metaphors' as 'schemata' or *clichés* that are practical adaptations, blocking access to properly philosophical 'speculation' (see Deleuze 1988; Bergson 1994). It is also noteworthy that Deleuze is aware of the technological background of the term '*cliché*' (see McLuhan 1970):

> Now this is what a *cliché* is. … We do not perceive the thing or the image in its entirety, we always perceive less of it, we perceive only what we are interested in perceiving, or rather what it is in our interest to perceive, by virtue of economic interests, ideological beliefs and psychological demands. We therefore normally only perceive in *clichés*. But if our … schemata jam or break, then a different type of image can appear. (Deleuze 2005a: 19–20)

21 Consider, most obviously, how 'artificial processes and systems' are taken up in multiple speculative and allegorical senses in filmic and written works of science fiction (see Hayles 1999; Kang 2011; Harlan et al. 2009).

Chapter three

1 This issue is particularly important in cases involving embodiment conditions concerning gender, ethnicity and disability (on this, see the end of Part 1, and Liu 2010; Hayles 1999; Galloway 2004: 184–96; Malabou 2009).

2 Hansen uses the term 'new media' to cover digitized information (including photographs, text, videos, and sound files) and its material technological conditions of possibility, including, but not limited to, the internet, smartphones, and virtual and 'mixed' reality technologies (Hansen 2004: 22, 2006: 8–9). For canonical remarks on the distinctions between old and new media, see Manovich (2001). On 'twenty-first-century media', see Hansen (2015). Hansen takes new media and 'twenty-first-century media' to constitute qualitative shifts in our understanding of technology (see Part 2). My argument, to be developed in this chapter and the following one, is that the concept of exceptional technologies is more nuanced for understanding the specificity of qualitative shifts in our understanding of technologies.

3   On these complexities, see Castells (2010) and Galloway and Thacker (2007).
4   Respectively, the material 'networks of networks' of the net and the hypertext information space of the 'web' that runs through this infrastructure (see Galloway 2004: 29–53).
5   There is a sense in which all of these devices are, of course, always already 'networked' in Heidegger's philosophy, in a sense consistent with his holism (Heidegger 2005). A similar point holds from the perspective of Latour's approach to 'networks' (see Harman 2010).
6   On this, see Smith (2015).
7   See, in particular, Merleau-Ponty's broad and avowedly ambiguous sense of the 'body-schema' (1976).
8   Crucially, such an approach need not undermine the specificity of Dreyfus's commitments. On the contrary, by drawing out the sense in which these commitments are contingent and nested within further sets of (for instance historical, material, and evolutionary) conditions, it has the capacity to make his commitment to them all the more genuinely existentialist, in a political sense (on this, see Sartre 1960).
9   This position is distinct from 'Internet exceptionalism'. See the conclusion to this book, and Wu (2010).
10 See Bostrom (2005), Hansen (2006), Galloway (2012: 120–43) and Malabou (2004).
11 For an important critical contrast with the quasi-Hegelianism of Hansen's third point, see Danto (1986).
12 Hansen has explicitly characterized his approach as 'transcendental' in subsequent work (2006: 8–9, 39–41).
13 See also Hansen's work on 'twenty-first-century media' (2015: 3–4).
14 See, for example, Hayles's consideration of telegraphic technologies (2012: 123–70).
15 Such as discussions of the petrol engine in Simondon (2012a: 20–6), or the example of the blind man's cane in Merleau-Ponty (1976).
16 See also Deleuze's reference to 'a new conception of the transcendental' in Simondon (2004b: 124).
17 The problem Hansen highlights is an antinomy, in Kant's sense (Kant 2000: 459–550). Noting this does not commit me to a species of Kant's transcendental idealism in references to the 'transcendental' in what follows.
18 By 'transcendental' here, I mean 'having to do with the conditions for sense'. By 'empirical', I mean 'having to do with the (actual and possible) objects of sense'.
19 Hansen's elevation of new media art could, of course, be written off as a contingent matter of 'taste'. It should be noted, in this respect, that Hansen's more recent work has moved beyond the restrictions of this strategic decision (see Hansen 2015).
20 Both Hayles and Hansen draw on work in the 4e context (see Hayles 2012; Hansen 2015).

21 For a notorious defence of internalism against the extended mind thesis, see Adams and Aizawa (2010).

22 Heidegger features as a background influence for much work in the 4e context. However, 4e research has a considerably different tone and focus. For instance, it makes extensive use of examples from robotics and dynamic systems research (see Wheeler 2005, 2012; Clark 2011: 217).

23 Gallagher is also vocal on the theme (2012). For other approaches, see, for instance, Clark (2011), Wheeler (2005), Haugeland (2000) and Noë (2015).

24 On the four canonical problems of the 'differences argument', the 'coupling-constitution fallacy', 'cognitive bloat', and the 'mark of the cognitive', see Rowlands 2010: 85–107. For a suggested taxonomy of 'contextual factors', 'enabling conditions', and 'constitutive processes', see De Jaegher et al. (2010).

25 It could, of course, be objected here that Clark and Chalmers's use of the term 'dropping' is contingent, and that it could be replaced with a more formal terminology. What this would overlook are the ways in which matters of degree and ambiguity condition the richness of debates and further work in the 4e context. Rather than aberrant features to be eliminated, they are in fact constitutive conditions of 4e work in a transcendental sense. I take this point to be consistent with Rowlands's remarks on the 'noneliminable' transcendental role of sense, as discussed above.

26 A classic recurrent example in the literature is that of 'Otto's notebook', as discussed by Clark and Chalmers (2011).

27 By 'methodological naturalism', I mean the view that philosophy is continuous with the methods and results of the natural sciences (see Reynolds 2014; Papineau 2016).

28 In principle, Rowlands's position in *The New Science* is, for instance, consistent with the thesis of the socially extended mind. In fact, however, he favours a more restricted epistemological purview than that of Gallagher and Crisafi (on this, see his worries on the problem of cognitive bloat, and on 'ownership' as a mark of the cognitive (2010: 107–62)).

29 Considered just in terms of cases covered in this chapter, an expanded 4e purview might, for instance, be used to further develop Dreyfus's discussion of the internet, Hayles's accounts of attentional processes in reading technologies (2012) and Hansen's approach to embodiment conditions in case studies of new media artworks (2000). In making this point, it should be noted that Hayles and Hansen already draw on 4e work (Hayles 2012; Hansen 2015). The claim I am making here is simply that there is scope to take this treatment further, both in work in media theory, and in the 4e context.

30 Without seeking to offer an exhaustive list, there are three areas for further work with striking crossover potentials: 1) *the nature and scope of case studies* (how might case studies developed in philosophy of technology and media theory be drawn upon to extend the 4e purview beyond what Gallagher calls its 'typical' examples, and how would this work the other way?); 2) *differing conceptions of the role of philosophy* (how might the tendency towards

critique in continentally influenced approaches in philosophy of technology and media theory inflect the 4e tendency to see the primary role of philosophy as one of conceptual clarification, and how would this work the other way?); 3) *a politics of the commons* (how might work in the 4e context be related to recent continentally influenced work on the theme of 'the commons'?) Of these, I take the third area to have the most exciting potential. As Matthew Crawford has noted in calling for an 'attentional commons' (2015: 11), the conditions of cognition on a 4e view can be situated in terms of 'the commons' (that is, resources that are shared by all, from material resources like water and air, to symbolic resources like language, ideas and open source code). As Crawford also notes, however, 'We do not [yet] have a political economy corresponding' to this (2015: 11; see also Dean 2012; Hardt and Negri 2004; Žižek 2017).

Chapter four

1 On this, see the general introduction to this book, n. 13.
2 A digested version appeared in September 1945 in the even more popular weekly magazine *Life*.
3 'As We May Think' was first published after the end of hostilities in Europe ('VE' day is 8 May), but before the end of hostilities in the Pacific theatre ('VJ' day is 2 September).
4 When I refer to 'networks' in this part, I am using the term in the sense of computer or data networks, and not in the broader sense implied, for instance, by Bruno Latour's 'Actor Network Theory' (Latour 2005). On the importance of networking, see Galloway and Thacker (2007).
5 On this concept, Paisley and Butler write:
> Scientists and technologists are guided by 'images of potentiality' – the untested theories, unanswered questions, and unbuilt devices that they view as their agenda for five years, ten years, and longer. (1977: 42)
6 For an overview of Bush's influence on these figures and others, see the panels assembled for the 1995 'Brown/MIT Vannevar Bush Symposium' (Brown/MIT 1995). See also Nyce and Kahn (1991: 136–7), Licklider (1960, 1968), Engelbart (1962), Nelson (1972), Houston and Harmon (2007: 66) and Shannon and Weaver (1962). For a critical perspective on Bush's putative influence, see Chun (2008).
7 The influence of 'As We May Think' on Tim Berners-Lee, inventor of the Web, is acknowledged but ambiguous (see Houston and Harmon 2007: 68).
8 See Houston and Harmon (2007: 59–61).
9 It is possible (but unnecessary) to go much further here. To cite only some of the most obvious issues: Bush's emphasis on the 'relative permanence' of stored items does not envisage challenges posed in archiving and information management by the scale of information generated in networked cultures, and by the relative impermanence of optical media and magnetic storage

devices; there is no sense of phenomena such as social media, 'information overload', or 'the curated self'; there is no sense of increased issues to do with data security and surveillance, as posed in networked cultures; sociologically, Bush's emphasis on centralized control overlooks the increased importance of sharing under conditions of 'symbolic capitalism', as well as the increased importance of avatars and passwords in constructing what Deleuze refers to as the 'dividual' (Deleuze 1995).

10 Methodological individualism in the social sciences is the doctrine that social phenomena must be explained with reference to how they result from individual actions (see Heath 2015).

11 Bush's emphasis on what Hayles has called the 'liberal subject' means that he has no appreciation of how, to cite Galloway, control exists after decentralization; that is, pervasively, through 'protocol' and distributed networks (Hayles 1999; Galloway 2012).

12 This is not to understate that there are also billions for whom networking has not become accessible as an 'everyday reality', for vexed political, economic and social reasons. Instead, it is to point towards the ways in which networking has changed conditions affecting both those included and those excluded (On issues to do with the 'digital divide', see Floridi (2014: 48–9)). For a more sceptical take on issues described in this paragraph, see Edgerton (2006).

13 Eugenics is the deeply controversial 'science of the hereditary improvement of the human race by selective breeding' (Bulmer 2003: 79). It was promoted by Galton throughout his career, and was given its official name by his 1883 book, *Inquiries into Human Faculty* (Galton 1907).

14 Galton was a cousin of Charles Darwin, was independently wealthy, and deeply politically and socially conservative (Bulmer 2003: 39). It would therefore be tempting to take him for an isolated caricature of the Victorian amateur scientist, and to write off his work as 'crank' or anachronistic. This, however, would be a mistake. What it would overlook is the extent of Galton's connections with the British scientific establishment, as well as influential contributions he made to various fields. Galton was, for instance, a vocal member of the British Association (BA) and, with others including Thomas Huxley and Herbert Spencer, a founding member of *The Reader* (the precursor to *Nature*) (Bulmer 2003: 37). In 1884, he served as consultant to the British Medical Association (BMA) for the production of a *Life History Album*, marketed to parents as a photographic record of the mental and physical developments of their children (Galton 1902). Further, Galton was, in an 1892 book, the first to systematize William Herschel's practice of fingerprinting as a criminological tool. Galton should therefore be read less a caricature of the scientist of his day, and more as a condensation of trends running through the Victorian scientific establishment, and his approach to composite photography is consistent with this picture. Although it coincided with a period of immense experimentation in photography, it was by no means 'hobbyist'. Instead, it built on prior innovative work Galton had conducted in anthropometric photography (Sera-Shriar 2015: 166–7), and he was sufficiently well-connected to gain access to exclusive photographic records as materials, including

portraits of hundreds of convicts held in Pentonville and Millbank prisons (Green 1985: 11).

15 On circularity, see Ellenbogen (2012: 121). On the nominalist/realist distinction, see Sekula (1986: 18) and Ellenbogen (2012: 109). On ideology, see Sekula (1986), Green (1985) and Bailey (2012).

16 Galton claimed his results were replicable in at least three senses. First, he claimed that ordering the component photographs differently produced consistent composites (1879a: 135). Second, he claimed that magic lanterns could be used to superimpose the images in a different way and to corroborate the results (Galton 1879b). Third, Galton published his method (Green 1985; Wade 2016).

17 This desire for exclusivity is consistent with Galton's claim that his results were replicable. He merely took his results to be replicable for a narrow elite, as is further consistent with his general ideological outlook.

18 In a general late-nineteenth-century context of widespread photographic realism, this was perhaps the broadest persuasive factor underpinning Galton's practice (Bailey 2015).

19 An 'Index', for Peirce, is 'a sign which refers to the Object. ... It denotes by virtue of being really affected by that Object' (Peirce 1955: 102). A 'Symbol', in contrast, is 'a sign which refers to the Object that it denotes by virtue of a law … which operates to cause the Symbol to be interpreted as referring to that Object' (1955: 102–03).

20 In comparing his work to that of artists such as Joshua Reynolds, Galton recognized as much (Ellenbogen 2012: 129–54).

21 The genetic fallacy is a fallacy of irrelevance, where the origins of something are taken to determine its contemporary worth.

22 Two famous figures influenced by this tear were Freud (1999: 224–5) and Wittgenstein (1965), who both take up composite photography as a figure for advances they were seeking to bring about in their respective work on dreams and the philosophy of language.

23 Today, we might add popular face morphing apps to this list, such as FaceApp and Instaface.

24 It should also be stressed that there is a sense in which Galton's approach turns out to be very Platonic indeed, in spite of the initial 'empiricist' premises noted above. In Platonic fashion, Galton seems to have arrived at his equivalences as different ways of explaining the same results to different audiences: one based on idea formation (perhaps for the lay person or psychologists), another based on art practice (for aesthetes), and another based on the principles of statistics (for statisticians and mathematicians) (see Ellenbogen 2012).

25 Ganson states:
> When I'm making these pieces, I'm always trying to find a point where I'm saying something very clearly and it's very simple, but also at the same time it's very ambiguous. And I think there's a point between simplicity and ambiguity which can allow a viewer to perhaps take something from it (2004).

26 Galton's composite photography also had an impossible aim, but it is more accurate to describe it as a failed technological practice: whereas impossibility is a marker of the success of 'Machine with Concrete', Galton did not recognize the impossibility of his aim (at least initially, and perhaps not at all (see Ellenbogen 2012)).

27 Ganson's own reading of the work emphasizes an existential paradox of 'stillness' at one end, and immense activity at the other (Ganson 2009).

28 More precisely: 2.191 trillion years (Blume 1998).

29 Including, for instance, slowness in contrast to immediacy, and the elevation of low-tech and analogue features in a contemporary context of design values focused on networked ICTs (Zeleny 2005).

30 In contrast to 'matters of fact', which Latour takes to be a modernist theoretical construction implicating fictions of detached 'subjects' and 'objects' (2014).

31 The second iteration of 'Machine with Concrete' has been on loan at San Francisco's Exploratorium museum since 2013. Ganson has been an artist in residency at the Exploratorium, which was founded by Frank Oppenheimer, and which is described as '[not] just a museum [but] an ongoing exploration of science, art and human perception' (Exploratorium 2017). 'Beholding the Big Bang' is displayed as part of the 'Gestural Engineering' exhibition that has been dedicated to Ganson's work at MIT Museum since 1995 (MIT 2017), and it formed part of the 'Imagining Deep Time' exhibition at the non-profit National Academy of the Sciences, August 2014–January 2015 (Talasek 2014).

32 For instance: by investigating case studies of exceptional technologies arrived at by inventors across the gender spectrum, by looking to examples from non-Western cultures, and by investigating exceptional technologies produced by research teams (in fields like engineering, architecture and the contemporary biomedical sciences).

Chapter five

1 If abstract myths and fictions have 'no clear relation', then they have a relation that is a clear candidate for philosophical clarification. If what Kroes and Meijers really mean is that they clearly have no relation, then this presupposes as clear a relation as can be given between two sets of entities (namely, a relation of no relation). Both of these issues are discussed in further detail below.

2 And this in spite of the fact that Kroes and Meijers's juxtaposed references to 'abstract myths and fictions' and 'engineering practice' imply that there are at least some sets of entities and practices that exist in relation to 'the real world of technology', but that are not transparently available to common sense.

3 By virtue of the priority Kroes and Meijers assign to the 'engineering sciences', it is also implied that philosophy of technology should turn away from a rich set of crossover potentials with other areas of work, such as software studies, art and design, cognitive science, and media theory.

4   Worse still, it makes it seem like some form of caricature of a thoroughly 'modern' Hegelian Dialectic, instead of pointing towards something like Latour's 'amodern' approach. The deeper fault here, however, may lie with Latour himself, who describes his own approach in terms of a picture of turning in the essay cited by Verbeek (Latour 1992a). The aim of Latour's 'amodern' approach is to make us aware of the extent to which 'we have never been modern'. By this, he means that we should challenge the epistemological picture laid down in philosophy since Kant, and developed in various ways by thinkers including Hegel, Heidegger, Habermas, Baudrillard and Lyotard (1993: 49–88). Against this picture, Latour claims we need to open up a 'second dimension' pertaining to all the issues that act as the repressed conditions for this 'modern' epistemological picture (1992a: 13). But why does Latour not realize the extent to which 'turning' remains a thoroughly 'modernist' picture of method (witness Kant's 'Copernican turn' and Heidegger's *Kehre*)? And why does he not realize that his own focus on conditions parallels something deeper and more interesting in the 'modernist' conception of method: namely, a sense of the transcendental as ongoing engagement with conditions? At the very least, Latour's description of his approach in terms of 'one more turn' amounts to an awkward shoehorning, as well as a bad precedent for thinkers like Verbeek to follow.

5   Bryant et al.'s recognition that 'it is difficult to find a single name' is a considerable understatement. In fact, speculative developments in recent continental philosophy are subject to even more pronounced tendencies towards fracture and specialization than those discussed in the case of philosophy of technology's empirical turn (see Zahavi 2016: 304). This is due, in part, to the encouragement of strong polemic across many of the recent speculative approaches (see, notoriously, Brassier's polemic against the term 'speculative realism' (Brassier and Rychter 2011)).

6   Given Harman's reference to it in the above passage, let me hazard an observation on Kant's 'Copernican Revolution' here (which is sometimes referred to as his 'Copernican turn'). In the *Critique of Pure Reason*, Kant famously states:

> Up to now it has been assumed that all our cognition must conform to the objects; but all attempts to find out something about them *a priori* through concepts that would extend our cognition have, on this presupposition, come to nothing. Hence let us once try whether we do not get farther with the problems of metaphysics by assuming that the objects must conform to our cognition, which would agree better with the requested possibility of an *a priori* cognition of them, which is to establish something about objects before they are given to us. This would be just like the first thoughts of Copernicus, who, when he did not make good progress in the explanation of the celestial motions if he assumed that the entire celestial host revolves around the observer, tried to see if he might not have greater success if he made the observer revolve and left the stars at rest. (2000: Bxvi)

Note that Kant forwards the transcendental idealist postulate that 'objects must conform to our cognition' as a 'presupposition' or 'assumption' here. My contention, developed over the course of this book, is that this points to an underlying sense of the transcendental as an approach to argument or method (that involves proceeding on the basis of an examined approach to presuppositions or conditions) that acts as a prior condition for the possibility of the anthropocentric metaphysical doctrine known as 'transcendental idealism'.

7. Kroes writes:

    > All the various forms of technology ... may hang together through ... family resemblance, without there being a common core element. Modern technology is a historically grown, highly complex and diverse phenomenon, a fact that should not be ignored by philosophers of technology. This is only possible through a shift from a global to a more local level of analysis. The richness of technology will become visible only by looking at modern technology through a magnifying glass. (2000: 28)

    The initial point against the reification of 'Technology' here is, I take it, very well-articulated, but the subsequent prescription in favour of an exclusively local focus is an overstatement at best, and, at worst, deeply misleading on philosophy of technology's capacity to also engage with more 'global' issues (for instance: the Anthropocene, the rise of automation, globalization, the implications of new ICTs for neoliberalism, and the growth of Information societies).

8. See, for instance, the way in which Galloway and Thacker develop a wide-ranging sense of 'networking' (2007).

9. Wittgenstein is especially suggestive on this point:

    > A main source of our failure to understand is that we don't have *an overview* of the use of our words. Our grammar is deficient in surveyability. A surveyable representation produces precisely that kind of understanding which consists in 'seeing connections'. Hence the importance of finding and inventing *intermediate links*. (2009: 54)

    Typically, a quote like this would be located in terms of a picture of Wittgenstein as a 'linguistic turn' thinker. But what if the emphasis fell on the technologically mediated picture of 'surveyability' or 'mapping' that he offers here? In what ways would this change our received picture of his method?

10. Although it has subsequently been instantiated as a literal institution in different ways, most obviously in penitentiary systems in the United States (see Brunon-Ernst 2012).

11. This point also applies to cases involving non-human animals: consider Harlow's famous experiments with Rhesus monkeys (2008), or to the development of mimicry techniques in nature, such as the case of the wasp and the orchid, as described by Deleuze (Deleuze and Parnet 2007: 2).

12. Foucault points to examples of barracks, poor houses, schools and hospitals (1991: 211–12).

13. On the limits in reading Foucault in this way, see Poster (2001a).

14. Consider, for instance, the concluding use of an 'anonymous text' from the social science journal *La Phalange* (1991: 307–08).

15. Crucially, Foucault's approach also has clearly criticizable limits that have formed the focus for subsequent philosophical work. Deleuze, for instance,

argues that an era of 'control societies' follows 'disciplinary society'. On this account, the appropriate speculative figure is not the panopticon, but 'information technology and computers', for the production of perpetually divided subjects (1995: 180). Galloway takes this Deleuzian approach further, arguing that '[computer] protocol is to control societies as the panopticon is to disciplinary society' (2004: 13). Poster, in contrast, has argued that that the figure of a 'super-Panopticon' best describes social conditions under late capitalism (2001a: 43–4).

16 What underlies Bryant's move is an attachment to Deleuze and Guattari's approach to 'desiring machines' in *Anti-Oedipus*. But Bryant decontextualizes Deleuze and Guattari's ontologizing move from the context of its development: as a polemical move against Freud and Lacan (2004b).

17 To cite an example of this from Bryant's book:

> Not all machines are material in nature. … A national constitution is not a being composed of fixed material parts like a cell phone, but is nonetheless a machine. A recipe does not itself have any ingredients, but is still a machine for operating on ingredients. A novel does not itself contain any people, rocks, heaths, animals, bombs, or airborne toxic events but nonetheless acts on other machines such as people, institutions, economies, etc. in all sorts of ways. Debt is nothing that we could identify as a material thing in the world, but is a machine that organizes the lives of billions of people. (Bryant 2014: 16)

The speculative instances covered in this litany are, of themselves, by no means uninteresting. The problem is that they emerge as distracted, forced and superficial where seemingly their *only* function in Bryant's book is to feature as instances of a litany.

18 Without wishing to resort to a litany, note that the panopticon can easily act as a provocation to consider cases involving non-human animals and entities: How does it relate to issues concerning zoos and animal welfare? What about industrialized farming? What about the growing impacts of automation and surveillance of human and non-human entities in factories and delivery centres? How does it compare to the functions of an insect colony?

Conclusion

1 'Technological exceptionalism' is a concept drawn from cyberlaw, where it refers to a form of technological determinism, where 'dramatic technological change necessitates systematic legal change' (Jones 2017: 41). A subspecies also discussed in this field is 'Internet exceptionalism' (Wu 2010). In political philosophy and economics, it is common to find references to notions of US or Chinese 'exceptionalism' (see, for instance, Galloway and Thacker 2007: 1–22). In recent speculative metaphysics, anthropocentric positions like Kant's transcendental idealism are often framed as forms of 'human exceptionalism' (Bryant 2014: 285; Bennett 2010: 34–7). I use the term 'technological exceptionalism' in preference to 'technological determinism' here, but their dangers are clearly related.

2 What counts as 'exceptional' is context-dependent. It would be absurd to privilege a 'memex' over a contemporary smartphone as an exceptional way of connecting to the internet; however, it might make more sense to privilege the former over the latter as a focal point for drawing out and challenging issues concerning cultures of networking in certain contexts.

3 What a developed sense of the transcendental ultimately teaches, I hold, is to focus critically and dynamically on the conditions that are constitutive of sense in a given context, and to be attuned to what its apparent exceptions can teach us. This holds for the local (for instance, the situation of a particular lab or design problem), up to the level of the global. One of the profound advantages of a developed sense of the transcendental, as such, is that it allows for engagement with how emergent englobing ways of making sense change how sense is made in other contexts (for instance, the Anthropocene, Information, the commons, neoliberalism).

4 Deleuze and Guattari famously defined philosophy as 'the art of forming, inventing, and fabricating concepts' (1994: 2).

5 It would, for instance, be quite easy to conceive of Ganson's *Machine with Concrete* as a 'quasi-object' in Latour's sense, but it would be difficult, unnecessary or unjustified to attempt to conceive of it as a work of new media, or as a form of pseudoscience.

REFERENCES

Achterhuis, H. (2001), *American Philosophy of Technology: The Empirical Turn*, trans. R. P. Crease, Bloomington: Indiana University Press.
Adams, F. and K. Aizawa (2010), 'Defending the Bounds of Cognition', in R. Menary (ed.), *The Extended Mind*, 67–80, Cambridge, MA: MIT Press.
Agamben, G. (2005), *State of Exception*, trans. K. Attell, Chicago: University of Chicago Press.
Aydin, C. and P. P. Verbeek (2015), 'Transcendence in Technology', *Techné*, 19: 314–57.
Bachelard, G. (1970), *Le Droit de rêver*, Paris: PUF.
Badiou, A. (1999), *Manifesto for Philosophy*, trans. N. Madarasz, Albany: SUNY Press.
Bailey, S. (2012), 'Francis Galton's Face Project: Morphing the Victorian Human', *Photography and Culture*, 5 (2): 189–214. doi:10.2752/1751452 12X13330132507077.
Battelle, J. (2005), *The Search: How Google and Its Rivals Rewrote the Rules of Business and Transformed Our Culture*, London: Nicholas Brealey Publishing.
Beiser, F. (2002), *German Idealism: The Struggle Against Subjectivism, 1781-1801*, Cambridge, MA: Harvard University Press.
Bell, D. (1991), *Husserl*, London: Routledge.
Bennett, J. (2010), *Vibrant Matter: A Political Ecology of Things*, Durham: Duke University Press.
Bennington, G. (2002), *Interrupting Derrida*, London: Routledge.
Bentham, J. (1995), *The Panopticon Writings*, M. Bozovic (ed.), London: Verso.
Berg Olsen, J. K. and E. Selinger, eds (2007), *Philosophy of Technology: Five Questions*, Copenhagen: Vince Inc. Press.
Berg Olsen, J. K., E. Selinger and S. Riis, eds (2009), *New Waves in Philosophy of Technology*, Basingstoke: Palgrave Macmillan.
Bergson, H. (1994), *L'Evolution créatrice*, Paris: PUF.
Bijker, W. E., T. P. Hughes and T. Pinch, eds (1987), *The Social Construction of Technological Systems: New Directions in the Sociology and History of Technology*, Cambridge, MA: MIT Press.
Blattner, W. (2007), 'Ontology, the A Priori, and the Primacy of Practice: An Aporia in Heidegger's Early Philosophy', in S. Crowell and J. Malpas (eds), *Transcendental Heidegger*, 10–27, Stanford: Stanford University Press.
Bloor, D. (1976), *Knowledge and Social Imagery*, London: Routledge & Kegan Paul.
Blume, H. (1998), 'Subtle Mechanisms', *The Atlantic Magazine*, 13 August 1998. Available online: https://www.theatlantic.com/past/docs/unbound/criticaleye/ce980813.htm (accessed 22 August 2017).

Bolter, J. D. (2000), *Writing Space: Computers, Hypertext, and the Remediation of Print*, Manwah: Erlbaum.
Bolter, J. D. and R. Grusin (1999), *Remediation: Understanding New Media*, Cambridge, MA: MIT Press.
Borgmann, A. (1984), *Technology and the Character of Everyday Life*, Chicago: University of Chicago Press.
Bostrom, N. (2005), 'In Defense of Posthuman Dignity', *Bioethics*, 19 (3): 202–14.
Brassier, R. (2007), *Nihil Unbound: Enlightenment and Extinction*, Basingstoke: Palgrave Macmillan.
Brassier, R. (2011), 'Concepts and Objects', in L. Bryant, N. Srnicek and G. Harman (eds), *The Speculative Turn: Continental Materialism and Realism*, 47–65, Victoria: re.press.
Brassier, R. and M. Rychter (2011), 'I Am a Nihilist Because I Still Believe in Truth', *Kronos* 1 (16). Available online: http://kronos.org.pl/numery/kronos-1-162011/ (accessed 8 March 2018).
Braver, L. (2007), *A Thing of This World: A History of Continental Anti-Realism*, Evanston: Northwestern University Press.
Braver, L. (2013), 'On Not Settling the Issue of Realism', *Speculations* (IV): 9–14.
Brey, P. (2008), 'Technology and Everything of Value', University of Twente Inaugural Speech Series. Available online: http://www.utwente.nl/gw/wijsb/organization/brey/Publicaties_Brey/Brey_2008_Oratie-ENG.pdf (accessed 22 August 2017).
Brey, P. (2010), 'Philosophy of Technology after the Empirical Turn', *Techné*, 14 (1): 36–48.
Brey, P. (2016), 'Constructive Philosophy of Technology and Responsible Innovation', in M. Franssen, P. Vermaas, P. Kroes and A. Meijers, (eds), *Philosophy of Technology after the Empirical Turn*, Philosophy of Engineering and Technology, 23, 127–43, Cham: Springer International.
Briggle, A. (2016), 'The Policy Turn in the Philosophy of Technology', in M. Franssen, P. Vermaas, P. Kroes and A. Meijers (eds), *Philosophy of Technology after the Empirical Turn*, Philosophy of Engineering and Technology, 23, 167–75, Cham: Springer International.
Brown, N. (2013), 'The Nadir of OOO: From Graham Harman's *Tool-Being* to Timothy Morton's *Realist Magic: Objects, Ontology, Causality*', *Parrhesia*, (17): 62–71. Available online: https://www.parrhesiajournal.org/parrhesia17/parrhesia17_brown.pdf (accessed 30 August 2017).
Brown/MIT. (1995), 'Brown/MIT Vannevar Bush Symposium'. Available online: http://www.cs.brown.edu/memex/Bush_Symposium.html (accessed 22 August 2017).
Brunon-Ernst, A. (2012), *Beyond Foucault: New Perspectives on Bentham's Panopticon*. Farnham: Ashgate.
Bryant, L. (2008), *Difference and Givenness: Deleuze's Transcendental Empiricism and the Ontology of Immanence*, Evanston: Northwestern University Press.
Bryant, L. (2014), *Onto-Cartography: An Ontology of Machines and Media*, Edinburgh: Edinburgh University Press.
Bryant, L., N. Srnicek and G. Harman (2011), 'Towards a Speculative Philosophy', in L. Bryant, N. Srnicek and G Harman (eds), *The Speculative Turn: Continental Materialism and Realism*, 1–19. Victoria: re.press.

Bulmer, M. (2003), *Francis Galton: Pioneer of Hereditary and Biometry*, Baltimore: John Hopkins University Press.
Bunge, M. (1985), *Treatise on Basic Philosophy (Vol 7, Part II): Life Science, Social Science and Technology*, Dordrecht: D. Reidel.
Bush, V. (1991a), 'Memex II', in J. M. Nyce and P. Kahn (eds), *From Memex to Hypertext: Vannevar Bush and the Mind's Machine*, 165–84, Boston: Academic Publishers.
Bush, V. (1991b), 'Memex Revisited', in J. M. Nyce and P. Kahn (eds), *From Memex to Hypertext: Vannevar Bush and the Mind's Machine*, 197–215, Boston: Academic Publishers.
Bush, V. (2017), 'As We May Think', *The Atlantic Magazine*, 1 July 1945. Available online: http://www.theatlantic.com/magazine/archive/1945/07/as-we-may-think/303881/ (accessed 22 August 2017).
Carr, N. (2015), *The Glass Cage: Where Automation Is Taking Us*, London: Penguin.
Carrouges, M. (1976), *LesMachines célibataries*, Paris: musée des arts décoratifs.
Casey, E. (1976), *Imagining: A Phenomenological Study*, Bloomington and London: Indiana University Press.
Cassirer, E. (2012), 'On Form and Technology', in A. S. Hoel and I. Folkvord (eds), *Ernst Cassirer on Form and Technology: Contemporary Readings*, 15–53, Basingstoke: Palgrave Macmillan.
Castells, M. (2010), *The Rise of the Network Society, vol.1: Economy, Society and Culture*, Oxford: Blackwell.
Cather, B. and B. Marsh (1997), 'Service Life Design of Concrete Structures', in E. A. Byers and T. McNulty (eds), *Management of Concrete Structures for Long-Term Serviceability*, 21–32, London: Thomas Telford Publishing.
Chamayou, G. (2013), *Théorie du drone*, Paris: La Fabrique.
Chase, J. and J. Reynolds (2011), *Analytic Versus Continental*, Durham: Acumen.
Chatonsky, G. (2013), 'La Solitude des machines'. Available online: http://chatonsky.net/folio/wp-content/uploads/2013/01/solitude-machines.pdf (accessed 22 August 2017).
Chun, W. H. K. (2008), 'The Enduring Ephemeral, or the Future Is a Memory', *Critical Inquiry*, 35 (1): 148–171.
Chun, W. H. K. (2011), *Programmed Visions: Software and Memory*, Cambridge, MA: MIT Press.
Clark, A. (1996), *Being There: Putting Brain, Body and World Together Again*, Cambridge, MA: MIT Press.
Clark, A. (2008), 'Pressing the Flesh: A Tension in the Study of the Embodied, Embedded Mind?', *Philosophy and Phenomenological Research*, 76 (1): 37–59.
Clark, A. (2011), *Supersizing the Mind: Embodiment, Action, and Cognitive Extension*, Oxford: OUP.
Clark, A. and D. Chalmers (2011), 'The Extended Mind', in A. Clark (ed.), *Supersizing the Mind*, 220–32, Oxford: Oxford University Press.
Coeckelbergh, M. (2017), *Using Words and Things: Language and Philosophy of Technology*, New York: Routledge.
Cohen, R. (2006), 'Technology: The Good, the Bad, and the Ugly', in E. Selinger (ed.), *Postphenomenology: A Critical Companion to Ihde*, 145–60, Albany: State University of New York Press.

Cowan, R. (1990), 'Nuclear Power Reactors: A Case Study in Technological Lock-In', *The Journal of Economic History*, 50 (3): 541–67.

Crawford, M. B. (2015), *The World Beyond Your Head: On Becoming an Individual in an Age of Distraction*, New York: Farrar Straus and Giroux.

Crowell, S. and J. Malpas, eds (2007), *Transcendental Heidegger*, Stanford: Stanford University Press.

Damasio, A. (2003), *Looking for Spinoza: Joy, Sorrow, and the Feeling Brain*, London: Harvest.

Danto. A. C. (1986), *The Philosophical Disenfranchisement of Art*, New York: Columbia University Press.

Davies, S. (2011), 'Still Building the Memex', *Communications of the ACM*, 54 (2): 80–8.

Dean, J. (2012), *The Communist Horizon*, London: Verso.

De Jaegher, H., E. Di Paolo and S. Gallagher (2010), 'Does Social Interaction Constitute Social Cognition?', *Trends in Cognitive Sciences*, 14 (10): 441–7.

Deleuze, G. (1988), *Bergsonism*, trans. H. Tomlinson and B. Habberjam, New York: Zone Books.

Deleuze, G. (1995), 'Postscript on Control Societies', in G. Deleuze (ed.), *Negotiations: 1972-1990*, trans. M. Joughin, 177–82, New York: Columbia University Press.

Deleuze, G. (2004a), *Difference and Repetition*, trans. P. Patton, London: Continuum.

Deleuze, G. (2004b), *Logic of Sense*, trans. M. Lester and C. Stivale and C. Boundas (ed.), London: Continuum.

Deleuze, G. (2005a), *Cinema 2: The Time-Image*, trans. H. Tomlinson and R. Galeta, London: Continuum.

Deleuze, G. (2005b), *Francis Bacon: The Logic of Sensation*, trans. D. W. Smith, London: Continuum.

Deleuze, G. and F. Guattari (1994), *What Is Philosophy?*, trans. G. Burchell and H. Tomlinson, London: Verso.

Deleuze, G. and F. Guattari (2004a), *A Thousand Plateaus*, trans. B. Massumi, London: Continuum.

Deleuze, G. and F. Guattari (2004b), *Anti-Oedipus*, trans. R. Hurley, M. Seem and H. R. Lane, London: Continuum.

Deleuze, G. and C. Parnet (2007), *Dialogues II*, trans. H. Tomlinson and B. Habberjam, New York: Columbia University Press.

Dennett, D. (2013), *Intuition Pumps and Other Tools for Thinking*, New York: Norton.

Derrida, J. (1972), 'La Mythologie blanche: la métaphore dans le texte philosophique', in J. Derrida (ed.), *Marges de la philosophie*, Paris: Presses Universitaires de France.

Derrida, J. (1973), *La Voix et le phénomène*, Paris: PUF.

Derrida, J. (1989), *Edmund Husserl's 'Origin of Geometry': An Introduction*, trans. J. P. Leavey, London: University of Nebraska Press.

Derrida, J. (2016), *Of Grammatology*, trans. G. C. Spivak, Baltimore: Johns Hopkins University Press.

Descartes, R. (2006), *Meditations, Objections and Replies*, trans. R. Ariew, Indianapolis: Hackett.

Descola, P. (2013), *Beyond Nature and Culture*, trans. J. Lloyd, Chicago: University of Chicago Press.
Dreyfus, H. (1991), *Being-In-The-World: Commentary on Heidegger's 'Being and Time', Division 1*, Cambridge, MA: MIT Press.
Dreyfus, H. (1992), *What Computers Can't Do: A Critique of Artificial Reason*, Cambridge, MA: MIT Press.
Dreyfus, H. (2001), *On the Internet*, London: Routledge.
Dummett, M. (1996), *Origins of Analytical Philosophy*, Cambridge, MA: Harvard University Press.
Dusek, V. (2006), *Philosophy of Technology: An Introduction*, Oxford: Blackwell.
Edgerton, R. (2006), *The Shock of the Old: Technology and Global History since 1900*, London: Profile Books.
Ellenbogen, J. (2012), *Reasoned and Unreasoned Images: The Photography of Bertillon, Galton, and Marey*, Pennsylvania: The Pennsylvania University Press.
Engelbart, D.C. (1962), 'Letter to Vannevar Bush and Program on Human Effectiveness', in J. M. Nyce and P. Kahn (eds), *From Memex to Hypertext: Vannevar Bush and the Mind's Machine*, 235–45, Boston: Academic Publishers.
Exploratorium. (2017), 'Exploratorium Website'. Available online: https://www.exploratorium.edu/ (accessed 29 August 2017).
Feenberg, A. (1999), *Questioning Technology*, London: Routledge.
Feenberg, A. (2002), *Transforming Technology: A Critical Theory Revisited*, Oxford: Oxford University Press.
Feenberg, A. (2005), *Heidegger and Marcuse: The Catastrophe and Redemption of History*, New York: Routledge.
Feenberg, A. (2009), 'Peter-Paul Verbeek: Review of What Things Do', *Human Studies*, 32 (2): 225–8. doi:http://dx.doi.org/10.1007/s10746-009-9115-3.
Feenberg, A. (2016), 'The Concept of Function in Critical Theory of Technology', in M. Franssen, P. Vermaas, P. Kroes and A. Meijers (eds), *Philosophy of Technology after the Empirical Turn*, Philosophy of Engineering and Technology, 23, 283–303, Cham: Springer International.
Floridi, L. (2013), *The Philosophy of Information*, Oxford: Oxford University Press.
Floridi, L. (2014), *The Fourth Revolution: How the Infosphere is Reshaping Human Reality*, Oxford: Oxford University Press.
Folkmann, M. N. (2013), *The Aesthetics of Imagination in Design*, Cambridge, MA: MIT Press.
Foucault, M. (1991), *Discipline and Punish: The Birth of the Prison*, trans. A. Sheridan, London: Penguin.
Foucault, M. (2002), *The Archaeology of Knowledge*, trans. A. M. Sheridan Smith, London: Continuum.
Franks, P. (1999), 'Transcendental Arguments, Reason, and Scepticism: Contemporary Debates and the Origins of Post-Kantianism', in R. Stern (ed.), *Transcendental Arguments: Problems and Prospects*, 111–45, Oxford: Clarendon.
Franssen, M. and S. Koller (2016), 'Philosophy of Technology as a Serious Branch of Philosophy: The Empirical Turn as a Starting Point', in M. Franssen, P. Vermaas, P. Kroes and A. Meijers (eds), *Philosophy of Technology after the Empirical Turn*, Philosophy of Engineering and Technology, 23, 31–61, Cham: Springer International.

Franssen, M., G. Lokhorst and I. van de Poel (2013), 'Philosophy of Technology', *The Stanford Encyclopedia of Philosophy* (Fall 2015 Edition), E. N. Zalta (ed.), Available online: http://plato.stanford.edu/archives/fall2015/entries/technology (accessed 22 August 2017).

Franssen, M., P. Vermaas, P. Kroes and A. Meijers (2016), 'Editorial Introduction: Putting the Empirical Turn into Perspective', in M. Franssen, P. Vermaas, P. Kroes and A. Meijers (eds), *Philosophy of Technology after the Empirical Turn*, Philosophy of Engineering and Technology, 23, 1–10, Cham: Springer International.

Freud, S. (1999), *The Interpretation of Dreams*, Oxford: Oxford University Press.

Friedman, M. (2002), 'Kant, Kuhn, and the Rationality of Science', *Philosophy of Science*, 69: 171–90.

Frieling, R. (2004), 'The Archive, the Media, the Map and the Text', Available online: http://www.medienkunstnetz.de/themes/mapping_and_text/archive_map/print/ (accessed 22 August 2017).

Friendly, M. (2008), 'The Golden Age of Statistical Graphics'. *Statistical Science*, 23 (4): 502–35.

Fuchs, C. (2011), *Foundations of Critical Media and Information Studies*, London: Routledge.

Fukuyama, F. (2002), *Our Posthuman Future: Consequences of the Biotech Revolution*, London: Profile Books.

Gadamer, H. (2004), *Truth and Method*, 2nd edn, trans. J. Weinsheimer and D. G. Marshall, London: Continuum.

Gallagher, S. (2012), *Phenomenology*, London: Palgrave Macmillan.

Gallagher, S. (2013), 'The Socially Extended Mind', *Cognitive Systems Research*, 25–26: 4–12. doi:https://doi.org/10.1016/j.cogsys.2013.03.008 (accessed 22 August 2017).

Gallagher, S. and A. Crisafi (2009), 'Mental Institutions', *Topoi*, 28 (1): 45–51. doi:10.1007/s11245-008-9045-0.

Galloway, A. (2004), *Protocol: How Control Exists after Decentralisation*, Cambridge, MA: MIT Press.

Galloway, A. (2012), *The Interface Effect*, Cambridge: Polity.

Galloway, A. and E. Thacker (2007), *The Exploit: A Theory of Networks*, Minneapolis: The University of Minnesota Press.

Galton, F. (1879a), 'Composite Portraits Made by Combining Those of Many Different Persons into a Single Figure', *Journal of the Anthropological Institute*, 8: 132–48.

Galton, F. (1879b), 'Generic Images', *Proceedings of the Royal Institution*, 9: 1–11.

Galton, F. (1882), 'Conventional Representation of the Horse in Motion', *Nature*, 26: 228–9.

Galton, F. (1900), 'Analytical Photography', *Photographic News*, 25: 135–8.

Galton, F. (1902), 'Life History Album', London: Macmillan.

Galton, F. (1907), *Inquiries into Human Faculty and Its Development*, 2nd edn, London: Dent.

Ganson, A. (2004), 'Moving Sculpture'. Available online: https://www.ted.com/talks/arthur_ganson_makes_moving_sculpture (accessed 22 August 2017).

Ganson, A. (2008), 'Machine with Concrete'. Available online: https://www.youtube.com/watch?v=5q-BH-tvxEg (accessed 22 August 2017).

Ganson, A. (2009), 'Machines and the Breath of Time'. Available online: http://longnow.org/seminars/02009/sep/14/machines-and-breath-time/ (accessed 22 August 2017).

Ganson, A. (2017), 'Machine with Concrete'. Available online: http://arthurganson.com/project/machine-with-concrete/ (accessed 22 August 2017).

Gardner, S. and M. Grist, eds (2015), *The Transcendental Turn*, Oxford: Oxford University Press.

Gates, K. (2011), *Our Biometric Future: Facial Recognition Technology and the Culture of Surveillance*, New York: New York University Press.

Gere, C. (2008), *Digital Culture*, 2nd edn, London: Reaktion.

Gleick, J. (2002), *Faster: The Acceleration of Just About Everything*, London: Little Brown.

Goddard K., A. Roudsari and J. C. Wyatt (2012), 'Automation Bias: A Systematic Review of Frequency, Effect Mediators, and Mitigators', *Journal of the American Medical Informatics Association: JAMIA*, 19 (1): 121–7. doi:10.1136/amiajnl-2011-000089.

Green, D. (1985), 'Veins of Resemblance: Photography and Eugenics', *The Oxford Art Journal*, 7 (2): 3–16.

Gurwitsch, A. (2010), *The Collected Works of Aron Gurwitsch (1901–1973)*, Volume III: The Field of Consciousness: Phenomenology of Theme, Thematic Field, and Marginal Consciousness. doi:10.1007/978-90-481-3346-8_1 (accessed 22 August 2017).

Hacking, I. (1990), *The Taming of Chance*, Cambridge: Cambridge University Press.

Hansen, M. B. N. (2004), *New Philosophy for New Media*, Cambridge, MA: MIT Press.

Hansen, M. B. N. (2006), *Bodies in Code: Interfaces with Digital Media*, London: Routledge.

Hansen, M. B. N. (2015), *Feed-Forward: On the Future of Twenty-First-Century Media*, Chicago: University of Chicago Press.

Hardt, M. and A. Negri (2004), *Multitude: War and Democracy in the Age of Empire*, London: Penguin.

Harlan, J., J. M. Struthers and C. Baker (2009), *AI: Artificial Intelligence. From Stanley Kubrick to Steven Spielberg: The Vision Behind the Film*, London: Thames and Hudson.

Harlow, H. (2008), 'The Monkey as a Psychological Subject', *Integrative Psychological and Behavioral Science*, 42 (4): 336–47. doi:10.1007/s12124-008-9058-7.

Harman, G. (2002), *Tool-Being: Heidegger and the Metaphysics of Objects*, Peru, Illinois: Open Court.

Harman, G. (2007), *Guerrilla Metaphysics: Phenomenology and the Carpentry of Things*, Peru, IL: Open Court.

Harman, G. (2010), 'Bruno Latour: King of Networks', in G. Harman (ed.), *Towards Speculative Realism: Essays and Lectures*, 67–92, Winchester: Zero Books.

Harman, G. (2011), *Towards Speculative Realism: Essays and Lectures*, Winchester: Zero Books.

Harris, P. and D. Lyon (2013), 'Catalyzing Dreams: Arthur Ganson's Mechanized Art', *Metalsmith*, 33 (2): 30–9.

Haugeland, J. (2000), *Having Thought: Essays in the Metaphysics of Mind*, Cambridge, MA: Harvard University Press.
Hayles, N. K. (1999), *How We Became Posthuman: Virtual bodies in Cybernetics, Literature, and Informatics*, Chicago: University of Chicago Press.
Hayles, N. K. (2012), *How We Think: Digital Media and Contemporary Technogenesis*, London: University of Chicago Press.
Heath, J. (2015), 'Methodological Individualism', *The Stanford Encyclopaedia of Philosophy* (Spring 2015 Edition), E. N. Zalta (ed.), Available online: https://plato.stanford.edu/archives/spr2015/entries/methodological-individualism (accessed 22 August 2017).
Hegel, G. W. F. (1977), *The Phenomenology of Spirit*, trans. A. V. Miller, Oxford: Clarendon.
Heidegger, M. (1977), *The Question Concerning Technology and Other Essays*, trans. W. Lovitt, New York: Harper and Row.
Heidegger, M. (1997), *Phenomenological Interpretation of Kant's 'Critique of Pure Reason'*, trans. P. Emad and K. May, Bloomington: Indiana University Press.
Heidegger, M. (2005), *Being and Time*, trans. J. Macquarrie and E. Robinson, Oxford: Blackwell.
Heidegger, M. (2010), '"Only a God Can Save Us": The *Spiegel* Interview (1966)', trans. W. J. Richardson, in T. Sheehan (ed.), *Heidegger: The Man and the Thinker*, 45–67, London: Transaction Publications.
Henry, M. (2003), *I Am the Truth: Toward a Philosophy of Christianity*, trans. S. Emanuel, Stanford: Stanford University Press.
Hillerbrand, R. and S. Roeser (2016), 'Towards a Third Practice Turn: An Inclusive and Empirically Informed Perspective on Risk', in M. Franssen, P. Vermaas, P. Kroes and A. Meijers (eds), *Philosophy of Technology after the Empirical Turn*, Philosophy of Engineering and Technology, 23, 145–66, Cham: Springer International.
Hintikka, J. (1972), 'Transcendental Arguments: Genuine and Spurious', *Nous*, 6 (3): 274–81.
Hookaway, C. (1999), 'Modest Transcendental Arguments and Sceptical Doubts: A Reply to Stroud', in R. Stern (ed.), *Transcendental Arguments: Problems and Prospects*, 173–88, Oxford: Clarendon.
Houston, R. D. and G. Harmon (2007), 'Vannevar Bush and Memex', *Annual Review of Information Science and Technology*, (41): 55–92.
Hui, Y. (2016), *The Question Concerning Technology in China: An Essay in Cosmotechnics*, Falmouth: Urbanomic.
Hume, D. (2007), *Enquiries Concerning Human Understanding and Concerning the Principles of Morals*, 3rd edn, Oxford: Clarendon.
Husserl, E. (1970), *The Crisis of European Sciences and Transcendental Phenomenology: An Introduction to Phenomenological Philosophy*, trans. D. Carr, Evanston: Northwestern University Press.
Husserl, E. (1988), *Cartesian Meditations: An Introduction to Phenomenology*, trans. D. Cairns, Dordrecht: Martinus Nijhoff Publishers.
Husserl, E. (1997), *Experience and Judgement*, trans. L. Landgrebe, Evanston: Northwestern University Press.
Husserl, E. (2002), *Ideas: General Introduction to Pure Phenomenology*, trans. W. R. Boyce Gibson, London: Routledge.

Ihde, D. (1990), *Technology and the Lifeworld: From Garden to Earth*, Bloomington: Indiana University Press.
Ihde, D. (2007), *Listening and Voice: Phenomenologies of Sound*, 2nd edn, Albany: SUNY Press.
Ihde, D. (2010), *Heidegger's Technologies: Postphenomenological Perspectives*, New York: Fordham University Press.
Ihde, D. (2012), *Experimental Phenomenology, Second Edition: Multistabilities*, Albany: SUNY Press.
Ihde, D. (2016), *Husserl's Missing Technologies*, New York: Fordham University Press.
James, I. (2012), *The New French Philosophy*, Cambridge: Polity Press.
James, W. and R. Perry (1922), *Essays in Radical Empiricism*, New York: Longmans, Green.
Jameson, F. (2002), *The Political Unconscious*, London: Routledge.
Jones, M. L. (2017), 'Does Technology Drive Law? The Dilemma of Technological Exceptionalism in Cyberlaw' (1 June 2017). Available online: SSRN:https://ssrn.com/abstract=2981855 (accessed 22 August 2017).
Kang, M. (2011), *Sublime Dreams of Living Machines: The Automaton in the European Imagination*, Cambridge, MA: Harvard University Press.
Kant, I. (1996), 'The Critique of Practical Reason', in I. Kant (ed.), *Practical Philosophy*, trans. M. Gregor, Cambridge: Cambridge University Press.
Kant, I. (2000), *The Critique of Pure Reason*, trans. P. Guyer and A. W. Wood, Cambridge: Cambridge University Press.
Kant, I. (2002), *Theoretical Philosophy after 1781*, H. Allison, P. Heath, G. Hatfield, M. Friedman, P. Guyer and A. Wood (eds), Cambridge: Cambridge University Press.
Kaplan, D. M. (2009a), 'Review: What Things Still Don't Do', *Human Studies*, 32 (2): 229–40.
Kaplan, D. M. (2009b), 'How to Read Technology Critically', in J. K. Berg Olsen, E. Selinger and S. Riis (eds), *New Waves in Philosophy of Technology*, 83–99, Basingstoke: Palgrave Macmillan.
Katz, B. M. (2011), 'Some Nonobject(ive) Reflections on the Nonobject', in B. Luckić (ed.), *Nonobject*, xxii–xxix, Cambridge, MA: MIT Press.
Körner, S. (1966), 'Transcendental Tendencies in Recent Philosophy', *The Journal of Philosophy*, 63 (19): 551–61.
Krippendorff, K. (2005), *The Semantic Turn: A New Foundation for Design*, London: Routledge.
Kroes, P. (2000), 'Engineering Design and the Empirical Turn in the Philosophy of Technology', in P. Kroes and A. Meijers (eds), *The Empirical Turn in the Philosophy of Technology*, 19–43, Bingley: Emerald Group.
Kroes, P. and A. Meijers (2000a), 'Introduction: A Discipline in Search of Its Identity', in P. Kroes and A. Meijers (eds), *The Empirical Turn in the Philosophy of Technology*, xvii–xxxv, Bingley: Emerald Group.
Kroes, P. and A. Meijers, eds (2000b), *The Empirical Turn in the Philosophy of Technology*, Bingley: Emerald Group.
Kroes, P. and A. Meijers (2016), 'Towards an Axiological Turn in the Philosophy of Technology', in M. Franssen, P. Vermaas, P. Kroes and A. Meijers (eds), *Philosophy of Technology after the Empirical Turn*, Philosophy of Engineering and Technology, 23, 11–30, Cham: Springer International.

Kuhn, T. (1962), *The Structure of Scientific Revolutions*, Chicago: University of Chicago Press.
Kuhn, T. (1981), 'A Function for Thought Experiments', in I. Hacking (ed.), *Scientific Revolutions*, 6–27, Oxford: Oxford University Press.
Kurzweil, R. (1999), *The Age of Spiritual Machines: When Computers Exceed Human Intelligence*, London: Penguin.
Lafont, C. (2007), 'Heidegger and the Synthetic A Priori', in S. Crowell and J. Malpas (eds), *Transcendental Heidegger*, 104–18, Stanford: Stanford University Press.
La Mettrie, J. (2006), *L'Homme-machine*, Paris: Gallimard.
Lanier, J. (2013), *Who Owns the Future?* London: Penguin.
Latour, B. (1992a), 'One More Turn After the Social Turn: Easing Science Studies into the Non-Modern World', in E. McMullin (ed.), *The Social Dimensions of Science*, 272–92, Notre Dame: Notre Dame University Press.
Latour, B. (1992b), 'Where Are the Missing Masses?', in W. Bijker and J. Law (eds), *Shaping Technology-Building Society. Studies in Sociotechnical Change*, 225–59, Cambridge, MA: MIT Press.
Latour, B. (1993), *We Have Never Been Modern*, trans. C. Porter, Cambridge, MA: Harvard University Press.
Latour, B. (2005), *Reassembling the Social: An Introduction to Actor-Network-Theory*, Oxford: Oxford University Press.
Latour, B. (2008), 'A Cautious Prometheus? A Few Steps Towards a Philosophy of Design (with Special Attention to Peter Sloterdijk)'. Available online: http://www.bruno-latour.fr/sites/default/files/112-DESIGN-CORNWALL-GB.pdf (accessed 23 August 2017).
Latour, B. (2014), 'Anthropology at the Time of the Anthropocene'. Available online: http://www.bruno-latour.fr/sites/default/files/139-AAA-Washington.pdf (accessed 23 August 2017).
Lee, P. (2004), *Chronophobia: On Time in the Art of the 1960s*, Cambridge, MA: MIT Press.
Licklider, J. C. R. (1960), 'Man Computer Symbiosis', *IRE Transactions on Human Factors in Electronics*, Vol. HFE-1, March 1960: 4–11. Available online: http://groups.csail.mit.edu/medg/people/psz/Licklider.html (accessed 23 August 2017).
Licklider, J. C. R. and R. W. Taylor (1968). 'The Computer as a Communication Device', *Science and Technology 1968*, Available online: http://gatekeeper.dec.com/pub/DEC/SRC/research-reports/abtrasts/src-rr-061.html (accessed 23 August 2017).
Liu, L. (2010), *The Freudian Robot: Digital Media and the Future of the Unconscious*, Chicago: University of Chicago Press.
Locke, J. (1993), *An Essay Concerning Human Understanding*, London: Everyman.
Luckić, B. (2011), *Nonobject*, Cambridge, MA: MIT Press.
McLuhan, M. (1970), *From Cliché to Archetype*, New York: Viking Press.
MacKay, R. and A. Avanessian, eds (2014), *#Accelerate: The Accelerationist Reader*, Falmouth: Urbanomic.
Malabou, C. (2004), *Que faire de notre cerveau?*, Paris: Bayard.
Malabou, C. (2005), *The Future of Hegel: Plasticity, Temporality and Dialectic*, London: Routledge.
Malabou, C. (2009), *Ontology of the Accident: An Essay on Destructive Plasticity*, trans. C. Shread, Cambridge: Polity.

Malabou, C. (2014), *Avant demain: Epigenèse et rationalité*, Paris: Presses Universitaires Français.
Malafouris, L. (2013), *How Things Shape the Mind: A Theory of Material Engagement*, Cambridge, MA: MIT Press.
Malpas, J. (1997), 'The Transcendental Circle', *Australasian Journal of Philosophy*, 75 (1): 1–20. doi:10.1080/00048409712347641.
Malpas, J. (2007), 'Heidegger's Topology of Being', in S. Crowell and J. Malpas (eds), *Transcendental Heidegger*, 119–34, Stanford: Stanford University Press.
Manovich, L. (2001), *The Language of New Media*, Cambridge, MA: MIT Press.
Marchant, G. E., B. R. Allenby and J. R. Herkert, eds (2011), *The Growing Gap Between Emerging Technologies and Legal-Ethical Oversight: The Pacing Problem*, Dordrecht: Springer Netherlands.
Mattes, E and F. Mattes (2001), 'Binneale.py'. Available online: http://0100101110101101.org/biennale-py/ (accessed 23 August 2017).
Maxwell, A. (2008), *Picture Imperfect: Photography and Eugenics*, Sussex: Sussex University Academic Press.
Meillassoux, Q. (2006), *Après la Finitude: un essai sur la nécessite de la contingence*, Paris: Seuil.
Merleau-Ponty, M. (1976), *Phénoménologie de la perception*, Paris: Gallimard.
Michael, M. (2006), *Technoscience and Everyday Life*, London: Open University Press.
MIT. (2017), 'Gestural Engineering: The Sculpture of Arthur Ganson'. Available online: https://mitmuseum.mit.edu/exhibition/gestural-engineering-sculpture-arthur-ganson (accessed 23 August 2017).
Mitcham, C. (1994), *Thinking through Technology: The Path between Engineering and Technology*, Chicago: University of Chicago Press.
Mitcham, C. (2002), 'Do Artifacts Have Dual Natures? Two Points of Commentary on the Delft Project', *Techné*, 6 (2): 9–12.
Moore, A. W. (2012), *The Evolution of Modern Metaphysics: Making Sense of Things*, Cambridge: Cambridge University Press.
Moran, D. (2000), *Introduction to Phenomenology*, London: Routledge.
Moran, D. (2007), 'Heidegger's Transcendental Phenomenology in Light of Husserl's Project of First Philosophy', in S. Crowell and J. Malpas (eds), *Transcendental Heidegger*, 135–51, Stanford: Stanford University Press.
Moran, D. (2012), *Husserl's 'Crisis of the European Sciences' and Transcendental Phenomenology: An Introduction*, Cambridge: Cambridge University Press.
Morozov, Y. (2013), *To Save Everything, Click Here: Technology, Solutionism and the Urge to Fix Problems That Don't Exist*, London: Allen Lane.
Mukherjee, S. (2017), 'A.I. Versus M.D.: What Happens When Diagnosis is Automated?', *The New Yorker*, 3 April 2017. Available online: http://www.newyorker.com/magazine/2017/04/03/ai-versus-md (accessed 23 August 2017).
Nagel, T. (1986), *The View From Nowhere*, Oxford: Oxford University Press.
Nelson, T. (1972), 'As We Will Think', in J. M. Nyce and P. Kahn (eds), *From Memex to Hypertext: Vannevar Bush and the Mind's Machine*, 245–61, Boston: Academic Publishers.
Noë, A. (2012), *Varieties of Presence*, Cambridge, MA: Harvard University Press.
Noë, A. (2015), *Strange Tools: Art and Human Nature*, New York: Hill and Wang.

NSA. (2003), 'Information Operations Roadmap'. Available online: http://nsarchive.gwu.edu/NSAEBB/NSAEBB177/info_ops_roadmap.pdf (accessed 23 August 2017).

Nyce, J. M. and P. Kahn (1991), 'The Idea of a Machine: The Later Memex Essays', in J. M. Nyce and P. Kahn (eds), *From Memex to Hypertext: Vannevar Bush and the Mind's Machine*, 113–44, Boston: Academic Publishers.

OED. (2017), 'paradox, n. and adj', *OED Online*. Available online: http://www.oed.com/view/Entry/137353?isAdvanced=false&result=1&rskey=KDxYMU& (accessed 23 August 2017).

Okrent, M. (2007), 'The "I Think" and the For-the-Sake-of-Which', in S. Crowell and J. Malpas (eds), *Transcendental Heidegger*, 151–68, Stanford: Stanford University Press.

Page, S. (2006), 'Path Dependence', *Quarterly Journal of Political Science*, 1 (1): 87–115. doi:http://dx.doi.org/10.1561/100.00000006.

Paisley, W. and M. Butler (1977), *Computer Assistance in Information Work*, Palo Alto, CA: Applied Communication Research.

Papineau, D. (2016), 'Naturalism', *The Stanford Encyclopedia of Philosophy* (Winter 2016 Edition), Edward N. Zalta (ed.), Available online: https://plato.stanford.edu/archives/win2016/entries/naturalism (accessed 23 August 2017).

Parikka, J. (2012), *What Is Media Archaeology?* Cambridge: Polity.

Pearl, S. (2010), *About Faces: Physiognomy in Nineteenth-Century Britain*, Cambridge, MA: Harvard University Press.

Peirce, C. S. (1955), *Philosophical Writings of Peirce*, J. Blucher (ed.), New York: Dover.

Pitt, J. C. (2011), *Doing Philosophy of Technology: Essays in a Pragmatist Spirit*, Dordrecht: Springer.

Pitt, J. C. (2016), 'The Future of Philosophy: A Manifesto', in M. Franssen, P. Vermaas, P. Kroes and A. Meijers (eds), *Philosophy of Technology after the Empirical Turn*, Philosophy of Engineering and Technology, 23, 83–92, Cham: Springer International.

Poster, M. (2001a), *The Information Subject*. Amsterdam: The Overseas Publishers Association.

Poster, M. (2001b), *What's the Matter with the Internet?* Minneapolis: University of Minnesota Press.

Poster, M. (2006), *Information Please: Culture and Politics in the Age of Digital Machines*, Durham, NC: Duke University Press.

Pinker, S. (2002), *The Blank Slate*, London: Penguin.

Priest, G. and F. Berto (2013), 'Dialetheism', *The Stanford Encyclopaedia of Philosophy* (Spring 2017 Edition), Edward N. Zalta (ed.), Available online: https://plato.stanford.edu/archives/spr2017/entries/dialetheism (accessed 23 August 2017).

Quetelet, A. (1846). *Lettres sur la théorie des probabilités, appliquée aux sciences morales et politiques*, Brussels: M. Hayez.

Reading, A. (2014), 'Seeing Red: A Political Economy of Digital Memory', *Media, Culture & Society*, 36 (6): 748–60. doi:10.1177/0163443714532980.

Reynolds, J. (2014), 'Transcendental Pragmatics? Deleuze, Pragmatism, and Metaphilosophy', in S. Bowden, S. Bignall and P. Patton (eds), *Deleuze and Pragmatism*, 235–51, London: Routledge.

Ricoeur, P. (1984), *Time and narrative. Vol.1*, trans. K. McLaughlin and D. Pellauer, Chicago: University of Chicago Press.
Rip, A. (2000), 'There's No Turn Like the Empirical Turn', in Kroes, P. and A. Meijers (eds), *The Empirical Turn in the Philosophy of Technology*, 3–17, Bingley: Emerald Group.
Rockmore, T. (2006), *In Kant's Wake: Philosophy in the Twentieth Century*, Oxford: Blackwell.
Roe Smith, M. and L. Marx, eds (1994), *Does Technology Drive History? The Dilemma of Technological Determinism*, Cambridge, MA: MIT Press.
Rowlands, M. (2010), *The New Science of the Mind: From Extended Mind to Embodied Phenomenology*, Cambridge, MA: MIT Press.
Rupert, R. (2004), 'Challenges to the Hypothesis of Extended Cognition', *Journal of Philosophy*, 101: 389–428.
Sartre, J. P. (1960), *Critique de la raison dialectique*, Paris: Gallimard.
Sartre, J. P. (1972), *The Transcendence of the Ego: An Existentialist Theory of Consciousness*, trans. F. Williams and R. Kirkpatrick, New York: Octagon.
Sartre, J. P. (2003), *Being and Nothingness: An Essay on Phenomenological Ontology*, trans. H. E. Barnes, London: Routledge.
Sartre, J. P. (2008), *La nausée*, Paris: Gallimard.
Sauvagnargues, A. (2008), *Deleuze: L'empirisme transcendantal*, Paris: Presses Universitaires de France.
Scharff, R. C. (2010), 'Technoscience Studies After Heidegger? Not Yet', *Philosophy Today*, 54.
Scharff, R. C. (2012), 'Empirical Technoscience Studies in a Comtean World: Too Much Concreteness?', *Philosophy & Technology*, 25 (2): 153–77. doi:10.1007/s13347-011-0047-2.
Scharff, R. C. and V. Dusek, eds (2014), *Philosophy of Technology: The Technological Condition*, 2nd edn, Oxford: Wiley-Blackwell.
Sekula, A. (1986), 'The Body and the Archive', *October*, (39): 3–64.
Sera-Shriar, E. (2015), 'Anthropometric Portraiture and Victorian Anthropology: Situating Francis Galton's Photographic Work in the Late 1870s', *History of Science*, 53 (2): 155–79.
Shannon, C. and W. Weaver (1962), *The Mathematical Theory of Communication*, Urbana: The University of Illinois Press.
Sicart, M. (2014), *Play Matters*, Cambridge, MA: MIT Press.
Simondon, G. (2012a), *Du mode d'existence des objets techniques*, Paris: Aubier.
Simondon, G. (2012b), 'On Techno-Aesthetics', trans. A. De Boever, *Parrhesia*, 12: 1–8. Available online: https://www.parrhesiajournal.org/parrhesia14/parrhesia14_simondon.pdf (accessed 23 August 2017).
Sloterdijk, P. (2005), 'Foreword to the Theory of Spheres', in M. Ohanian and J. C. Royoux (eds), *Cosmograms*, 223–41, New York: Lukas and Sternberg.
Smith, D. (2015), 'The Internet as Idea: For a Transcendental Philosophy of Technology', *Techné*, 19 (3): 381–410.
Smith, L. C. (1991), 'Memex as an Image of Potentiality Revisited', in J. M. Nyce and P. Kahn (eds), *From Memex to Hypertext: Vannevar Bush and the Mind's Machine*, 261–86, Boston: Academic Publishers.
Sparrow, T. (2014), *The End of Phenomenology: Metaphysics and the New Realism*, Edinburgh: Edinburgh University Press.

Srnicek, N. and A. Williams (2016), *Inventing the Future: Postcapitalism and a World Without Work*, London: Verso.

Stiegler, B. (1998), *Technics and Time, 1: The Fault of Epimetheus*, trans. R. Beardsworth and G. Collins, Stanford: Stanford University Press.

Stiegler, B. (2004), *Philosopher par accident: entretiens avec Elie During*, Paris: Editions Galilée.

Stiegler, B. (2010), *Ce qui fait que la vie vaut la peine d'être vécue*, Paris: Flammarion.

Stiegler, B. (2015), *La Societe automatique: 1. L'avenir du travail*, Paris: Fayard.

Stern, N. (2013), *Interactive Art and Embodiment: The Implicit Body as Performance*. Canterbury: Gylphi.

Stern, R., ed. (1999), *Transcendental Arguments: Problems and Prospects*, Oxford: Clarendon.

Strawson, P. (1959), *Individuals: An Essay in Descriptive Metaphysics*, London: Methuen and co.

Strawson, P. (1966), *The Bounds of Sense: An Essay on Kant's Critique of Pure Reason*, London: Methuen and co.

Stroud, B. (1968), 'Transcendental Arguments', *The Journal of Philosophy*, 65 (9): 241–56.

Stroud, B. (1999), 'The Goal of Transcendental Arguments', in R. Stern (ed.), *Transcendental Arguments: Problems and Prospects*, Oxford: Clarendon.

Swain, M. (1979), 'Justification and the Basis of Belief', in G. S. Pappas (ed.), *Justification and Knowledge: New Studies in Epistemology*, 25–49, Dordrecht: D. Reidel.

Talasek, J. D. (2014), 'Imagining Deep Time'. Available online: http://www.cpnas.org/exhibitions/imagining-deep-time-catalogue.pdf (accessed 23 August 17).

Taylor, C. M. (1995), *Philosophical Arguments*, Cambridge, MA: Harvard University Press.

Taylor, M. C. (2014), *Speed Limits: Where Time Went and Why We Have So Little Left*, London and New Haven: Yale University Press.

Thomson and Craighead (2014), *Stutterer*. Available online: http://www.ucl.ac.uk/slade/slide/stutterer.html (accessed 23 August 17).

Triclot, M. (2008), *Le Moment cybernétique: La constitution de la notion de l'information*, Paris: Editions Champ Vallon.

Vadivambal, R. and D. S. Jayas (2016), *Bio-Imaging: Principles, Techniques and Applications*, Boca Raton: Taylor and Francis.

Van den Eede, Y. (2013), *Amor Technologiae: Marshall McLuhan as Philosopher of Technology*, Brussels: VUB Press.

Varela, F., E. Rosch and E. Thompson (1991), *The Embodied Mind: Cognitive Science and Human Experience*, Cambridge, MA: MIT Press.

Verbeek, P. P. (2005), *What Things Do: Philosophical Reflections on Technology, Agency and Design*, Philadelphia: Penn State Press.

Verbeek, P. P. (2011), *Moralizing Technology: Understanding and Designing the Morality of Things*, Chicago: University of Chicago Press.

Vermaas, P. (2016), 'An Engineering Turn in Conceptual Analysis', in M. Franssen, P. Vermaas, P. Kroes and A. Meijers (eds), *Philosophy of Technology after the Empirical Turn*, Philosophy of Engineering and Technology, 23, 269–82, Cham: Springer International.

Virilio, P. (2006), *Speed and Politics*, trans. M. Polizzotti, Los Angeles: Semiotext(e).

Wade, N. J. (2016), 'Faces and Photography in 19th-Century Visual Science', *Perception*, 45 (9): 1008–35.
Watts, R., J. Bessant and R. Hil (2008), *International Criminology: A Critical Introduction*, New York: Routledge.
Watson, I. (2012), *The Universal Machine: From the Dawn of Computing to Digital Consciousness*, New York: Copernicus Books.
Wegenstein, B. (2010), 'Body', in W. J. T. Mitchell and M. B. N. Hansen (eds), *Critical Terms for Media Studies*, 19–34, Chicago: University of Chicago Press.
Weiser, M. (1991), 'The Computer for the 21st-Century', *Scientific American*, 265 (3): 94–104.
Wheeler, M. (2005), *Reconstructing the Cognitive World: The Next Step*, Cambridge, MA: MIT Press.
Wheeler, M. (2012), 'Minds, Things, and Materiality', in J. Schulkin (ed.), *Action, Perception, and the Brain: Adaptation and Cephalic Expression*, 147–63, Basingstoke: Palgrave-Macmillan.
Wiener, N. (1968), *The Human Use of Human Beings: Cybernetics and Society*, London: Sphere Books.
Williamson, T. (2007), *The Philosophy of Philosophy*, Oxford: Blackwell.
Wikipedia contributors. (2017), 'Turkish Archery', *Wikipedia*. Available online: https://en.wikipedia.org/w/index.php?title=Special:CiteThisPage&page=Turkish_archery&id=765125914 (Accessed 23 August 2017).
Wittgenstein, L. (1965), 'A Lecture on Ethics', *Philosophical Review*, 74 (1): 3–12.
Wittgenstein, L. (2009), *Philosophical Investigations*, trans. G. E. M. Anscombe and R. Rhees, Oxford: Wiley-Blackwell.
Wu, T. (2010), 'Is Internet Exceptionalism Dead?' in B. Szoka and A. Marcus (eds), *The Next Digital Decade: Essays on the Future of the Internet*, Washington DC: TechFreedom. Available online: https://ssrn.com/abstract=1752415 (accessed 23 August 2017).
Zahavi, D. (2003), *Husserl's Phenomenology*, Stanford: Stanford University Press.
Zahavi, D. (2008), *Subjectivity and Selfhood: Investigating the First-Person Perspective*, Cambridge, MA: MIT Press.
Zahavi, D. (2015), 'Husserl and the Transcendental', in S. Gardner and M. Grist (eds), *The Transcendental Turn*, 228–43, Oxford: Oxford University Press. doi:10.1093/acprof:oso/9780198724872.003.0011.
Zahavi, D. (2016), 'The End of What? Phenomenology vs. Speculative Realism', *International Journal of Philosophical Studies*, 24 (3): 289–309. doi:10.1080/09672559.2016.1175101.
Zeleny, M. (2005), *Human Systems Management: Integrating Knowledge, Management, and Systems*, Singapore: World Scientific Publishing Co.
Žižek, S. (2017), *The Courage of Hopelessness: Chronicles of a Year of Acting Dangerously*, London: Allen Lane.
Zweir, J., V. Blok and P. Lemmens (2016), 'Phenomenology and the Empirical Turn: A Phenomenology Analysis of Postphenomenology', *Philosophy & Technology*. doi:10.1007/s13347-016-0221-7.

INDEX

Note: Locators preceded by n. denotes note number

'4e' cognitive science 7, 56–7, 69–77, 131

Achterhuis, Hans 2, 12, 28–32, 110, 133–4 n.9
Anders, Günther 28
Arendt, Hannah 1, 28
Avant demain (Malabou) 7, 12–16, 135 n.5

Bailey, Suzanne 92
Beholding the Big Bang (Ganson) 98–104
Bennington, Geoffrey 20
Bentham, Jeremy 8, 121–4
'Big Tech' 139 n.16
blank page (case study of) 7, 35–53, 138 n.3, 139 n.11, 139 n.12
Brey, Philip 28–32, 113, 136 n.19
Bryant, Levi
  'onto-cartography' and 125–6
  speculative turn and 5, 107, 114–18, 147 n.5, 149 n.16, 149 n.17
  transcendental empiricism and 20, 136 n.17
Bush, Vannevar 5, 8, 77–86, 97, 130, 143 n.6, 143–4 n.9

'capital 'T'' Technology 2, 28–30, 112
Chase, James 135 n.2, 136 n.8
Chun, Wendy Hui Kyong 55, 67, 78, 84
Clark, Andy 7, 56, 70–4, 142 n.25
'classical' philosophy of technology 2–3, 12, 28–30, 32, 36, 109, 112, 130
Coeckelbergh, Mark 133 n.8

composite photography (Galton) 77, 86–97
continental philosophy 5–6, 8, 11–16, 39, 45–53, 65–9, 107–9, 114–19, 129–32, 133 n.7, 137 n.22
  'bad' tendencies of 13, 45
'correlationism' 14–15, 21, 22–3, 107, 115–19, 136 n.17
Crawford, Matthew B. 143 n.30
The Crisis of European Sciences and Transcendental Phenomenology (Husserl) 36–51, 138 n.5

deconstruction 13
Deleuze, Gilles
  blank page and 37, 48–51
  '*cliché*' and 50–1, 140 n.20
  embodiment conditions and 55
  philosophy of technology and 1, 143–4 n.9, 148–9 n.15
  transcendental and 13, 20, 26, 39, 47, 135 n.14, 136 n.17, 141 n.16
Deleuze, Giles, and Félix Guattari 118, 125–6, 149 n.16, 150 n.4
Derrida, Jacques 13–14, 20, 39, 47, 138 n.3
design 51–3, 101–5
'design fiction' 101–2, 130
determinism (technological) 2, 28, 123, 137 n.20, 149 n.1
Discipline and Punish (Foucault) 8, 109, 121–7
Dreyfus, Hubert
  'coping' and 21
  embodiment conditions and 55–63, 70, 75, 141 n.8

INDEX

Heidegger and 57–61
On the Internet 7, 57–63, 142 n.29
philosophy of technology and 133–4 n.9
transcendental and 7, 55–63
'drivers' 134 n.10

Ellenbogen, Josh 87–96, 145 n.24, 146 n.26
Ellul, Jacques 28
embodiment 7, 55–77
empirical turn 2, 6, 8, 27–33, 45–53, 78, 100, 107–14, 120, 124, 129–32, 133–4 n.9
empiricism 30, 90–1
engineering 51–3, 109–14
essentialism 12, 28
exam 43–4
'exceptionalism' (technological) 8, 105, 129–32, 149 n.1
'exceptional technologies' 1–9, 12, 28–33, 35–6, 42, 44, 46, 52–3, 55–7, 60–1, 65–9, 73–6, 77–105, 107, 109, 121–7, 129–32, 134 n.12, 134 n.13

Foucault, Michel
 'archaeology' and 65
 blank page and 43
 panopticon and 8, 109, 121–7
 philosophy of technology and 1, 5
 transcendental and 14, 20
Franks, Paul W. 17, 20, 23, 26
Franssen, Maarten 51–3
Franssen, Maarten, and Stefan Koller 37, 46–7
Freud, Sigmund 145 n.22
Fromm, Erich 28

Galileo, Galilei 40–1
Gallagher, Shaun 7
Gallagher, Shaun, and Anthony Crisafi 73–6
Galloway, Alexander R. 67, 84, 93, 97, 137 n.1, 139 n.16, 140 n.1, 141 n.4, 144 n.11, 148–9 n.15
Galloway, Alexander R., and Eugene Thacker 58, 68, 141 n.3, 143 n.4, 148 n.8, 149 n.1

Galton, Francis
 biographical note 144–5 n.14
 composite photography and 5, 8, 77–8, 86–97, 130, 145 n.16, 145 n.17, 145 n.24, 146 n.26
 eugenics and 144 n.13
Ganson, Arthur 5, 8, 78, 97–104, 145 n.25, 146 n.27, 146 n.31, 150 n.5
Gardner, Sebastian 16–18, 135–6 n.7

Hansen, Mark B. N.
 embodiment conditions and 7, 55–6, 64–70, 142 n.29
 media theory and 7, 55–6, 64–70, 75, 137 n.1
 transcendental and 141 n.12, 141 n.17
 'twenty-first-century media' and 97, 134 n.12, 140 n.2, 141 n.13
Harman, Graham
 Latour and 133 n.8, 141 n.5
 'object-oriented ontology' and 124–6
 speculative turn and 114–18
 transcendental and 147–8 n.6
Hayles, N. Katherine
 attention and 142 n.29
 blank page and 44
 embodiment conditions and 7, 45, 55–6, 62–70, 140 n.1
 liberal subject and 144 n.11
 media theory and 55–6, 62–70, 75, 137 n.1
 science fiction and 140 n.21
 telegraphic technologies and 141 n.14
Hegel, Georg W. F. 1, 13, 38
 transcendental and 20, 137 n.25, 138 n.4
Heidegger, Martin
 blank page and 37, 48–9
 phenomenology and 1, 13, 38–9, 48–9, 57, 72, 124–5
 philosophy of technology and 1, 12, 27–9, 58, 61, 110, 133 n.8, 141 n.5
 transcendental and 7, 11, 13, 20–7, 47, 72, 136 n.11, 138 n.5, 147 n.4
Hintikka, Jaakko 18–20

INDEX

historical materialism 31–2
historicising 31–3
Husserl, Edmund
  blank page and 36–51, 139 n.12, 139 n.14, 139 n.15
  embodiment conditions and 55
  imaginative variation and 7, 36–51, 131, 138 n.6
  phenomenology and 1, 13, 20, 36–51, 138 n.7
  philosophy of technology and 1, 36–51
  transcendental and 20, 36–51, 72, 138 n.4, 138 n.5

Ideas: General Introduction to Pure Phenomenology (Husserl) 36–51, 138 n.5
Ihde, Don
  empirical turn and 28, 31, 39
  Husserl and 36, 39–42, 47, 49, 139 n.9, 140 n.18
  philosophy of technology and 133–4 n.9, 136 n.19, 137 n.21, 137 n.22, 137 n.24
imaginative variation 7, 36–9, 41–53, 138 n.6
Internet 7, 56–61
  search engine 37, 43–4

Jaspers, Karl 28
Jonas, Hans 28

Kant, Immanuel
  continental philosophy and 2, 5–6, 11–16, 107–8, 114–19
  Heidegger and 23–7
  philosophy of technology and 1, 29
  transcendental and 2, 5–6, 7, 11–27, 36, 39, 48, 72, 107–8, 114–19, 135 n.2, 136 n.8, 136 n.15, 138 n.5, 147 n.4, 147–8 n.6
Kaplan, David M. 31–2
Katz, Barry M. 101–2
kinetic sculpture 102, 130
Kroes, Peter 120, 148 n.7
Kroes, Peter, and Anthonie Meijers 2, 32, 109–14, 133–4 n.9, 146 n.1, 146 n.2, 146 n.3

Latour, Bruno
  'Actor-Network Theory' and 133 n.8, 141 n.5, 143 n.4
  'amodern' approach of 147 n.4
  design and 102–4
  quasi-object and 130, 150 n.5
  Science and Technology Studies and 112–13, 133–4 n.9
  speculative turn and 115
Licklider, Joseph C.R. 82
linguistic turn 115–16, 148 n.9
Liu, Lydia 67, 140 n.1
Locke, John 35, 138 n.2

Machine with Concrete (Ganson) 5, 97–104, 150 n.5
McLuhan, Marshall 133 n.8
Malabou, Catherine
  continental philosophy and 7, 11–16, 19–21
  correlationism and 135 n.5
  embodiment conditions and 55, 140 n.1
  plasticity and 15
  transcendental and 7, 11–16, 19–21, 135 n.6
Malpas, Jeff
  transcendental and 12, 20, 23–7, 135–6 n.7
mapping (as method) 3, 8, 32–3, 109, 119–27, 129–32
Marcuse, Herbert 28
Marx, Karl 1, 31–2, 137 n.25
media archaeology 9, 65
media theory 62–76
Meillassoux, Quentin
  correlationism and 21–3, 118, 135 n.5
  speculative turn and 15, 115, 118
memex (Vannevar Bush) 77–86
Merleau-Ponty, Maurice 18, 39, 47, 55, 57, 66, 138 n.5, 141 n.7
Mitcham, Carl 1, 51
  'humanities philosophy of technology' and 55, 138 n.3
Moore, Adrian W. 135 n.4
Moran, Dermot 20
Mumford, Lewis 28

Nelson, Ted 82
neuroscience 11, 15, 61, 115
new media 64–76, 130, 140 n.2
Noë, Alva 55, 60, 142 n.23

Okrent, Mark 21–2
On the Internet (Dreyfus) 7, 57–61

'pacing problem' 134 n.10
panopticon 109, 121–7
'path dependence' 134 n.10
Peirce, Charles Sanders 90–1, 145 n.19
philosophy of technology 1–9, 27–33, 40, 44–53, 77–8, 107–14, 129–32, 148 n.7
picture of method 2, 3, 8, 12, 28–33, 107–27, 129–32, 148 n.9
Pinker, Stephen 139 n.12
Pitt, Joseph C. 137 n.24
Poster, Mark 55, 67, 148 n.13, 148–9 n.15
Priest, Graham, and Francesco Berto 134 n.11
problem of relevance 7, 35–6, 48–51

reification (fallacy of) 2, 12, 16, 27–30, 32, 61, 109–10, 130
remediation 35, 138 n.2
Reynolds, Jack 136 n.8
Ricoeur, Paul 14
'roadmaps' 134 n.10
Rowlands, Mark 7–8, 56, 69–73

Sartre, Jean Paul 37, 39, 47, 72, 100, 122, 138 n.5, 141 n.8
Sauvagnargues, Anne 20
Scharff, Robert C. 30–1, 133 n.8
Sekula, Allan 95–6
Shannon, Claude 82

Simondon, Gilbert 64–7, 141 n.15, 141 n.16
Sloterdijk, Peter 102
speculative turn 8, 11, 15, 114–19
Srnicek, Nick 114
Stern, Robert 18–23, 136 n.8
Stiegler, Bernard 67, 133, 139 n.10
Strawson, Peter 18, 136 n.8
Stroud, Barry 18–23, 136 n.8

Taylor, Charles 17–18, 23, 26
transcendental (philosophical sense of) 1–9, 11–33, 35–6, 47–51, 55–76, 85, 107–9, 129–32, 135 n.2, 150 n.3
transcendental arguments 17–19, 22, 23–7, 136 n.8
'transcendental idealism' (Kant's doctrine of) 14–15, 16–17, 19, 115–16, 147 n.6
'transcendental phenomenology' (Husserl's doctrine of) 36–51, 138 n.5

Van Den Eede, Yoni 133 n.8
Verbeek, Peter-Paul
  empirical turn and 12, 28–32, 100, 112–13, 133–4 n.9, 136 n.19
  Latour and 147 n.4
  transcendental and 28–32, 110

Williamson, Timothy 135 n.4
Wittgenstein, Ludwig 2, 44–5, 145 n.22, 148 n.9
Wu, Tim 141 n.9

Zahavi, Dan 20
'zeitgeist-seizing' technologies 31, 35, 135 n.15, 137 n.23
Žižek, Slavoj 115

www.ingramcontent.com/pod-product-compliance
Lightning Source LLC
Chambersburg PA
CBHW051812230426
43672CB00012B/2706